THE Chinese
ELECTRONICS
INDUSTRY

The Electronics Industry Research Series

- The Taiwan Electronics Industry
- The Singapore and Malaysia Electronics Industries
- The Korean Electronics Industry
- The Japanese Electronics Industry
- The Chinese Electronics Industry

THE Chinese
ELECTRONICS INDUSTRY

Michael Pecht

Chung-Shing Lee

Zong Xiang Fu

Jiang Jun Lu

Wang Yong Wen

 CRC Press
Taylor & Francis Group
Boca Raton London New York

CRC Press is an imprint of the
Taylor & Francis Group, an **informa** business

CRC Press
Taylor & Francis Group
6000 Broken Sound Parkway NW, Suite 300
Boca Raton, FL 33487-2742

© 1999 by Taylor & Francis Group, LLC
CRC Press is an imprint of Taylor & Francis Group, an Informa business

No claim to original U.S. Government works

ISBN 13: 978-0-8493-3174-9 (pbk)
ISBN 13: 978-1-138-43464-6 (hbk)

Visit the Taylor & Francis Web site at
http://www.taylorandfrancis.com

and the CRC Press Web site at
http://www.crcpress.com

AUTHORS

Michael G. Pecht is the Director of the CALCE Electronic Products and Systems Center at the University of Maryland and a Full Professor with a three way joint appointment in Mechanical Engineering, Engineering Research, and Systems Research. Dr. Pecht has a BS in Acoustics, an MS in Electrical Engineering and an MS and PhD in Engineering Mechanics from the University of Wisconsin. He is a Professional Engineer, an IEEE Fellow and an ASME Fellow. He has written fourteen books on electronics products development. He served as chief editor of the IEEE Transactions on Reliability for eight years and on the advisory board of IEEE Spectrum. He is currently the chief editor for *Microelectronics Reliability*, an associate editor for the *IEEE Transactions on Components, Packaging, and Manufacturing Technology*; and on the advisory board of the *Journal of Electronics Manufacturing*. He serves on the board of advisors for various companies and consults, providing expertise in strategic planning in the area of electronics products development and marketing.

Chung-Shing Lee is an Assistant Professor of Information Systems & Technology Management at the Pacific Lutheran University in Tacoma, Washington. Dr. Lee holds a B.A. degree from National Taiwan University, an M.A. degree in Economics from the University of Maryland, and a doctorate in Engineering and Technology Management from the George Washington University. Dr. Lee has many years of industrial consulting experience and was a Research Associate at the University of Maryland's CALCE Electronic Products and Systems Consortium. His research interests and publications are in the areas of Asian technology development and industry analysis, technology and innovation management, and strategic use of information technology.

Xiang-Fu Zong is the Chair Professor and Dean of the Technological Sciences and Engineering School at the Fudan University in Shanghai, The People's Republic of China.

Jun-Lu Jiang and Yong-Wen Wang are members of the State Science and Technology Commission, The People's Republic of China.

PREFACE

Since the open-door policy began in 1979, China has boasted one of the fastest growth economies in the world and one of the largest future markets. by the end of 1998, China had accumulated over US $144 billion of foreign exchange reserve, second only to Japan.

Although economic liberalization swept across all the sectors of the industrial economy, electronics and information technology have been targeted in particular for growth through export. In addition to simplifying the licensing and foreign investment policies, the government targeted technologies such as electronics for major development programs. Well-funded programs such as the "Torch Program" and the Economic & Technological Development Zones are functioning as national centers of excellence at the forefront of the electronics technology revolution. Recently, the Chinese central government is encouraging foreign investment in the so-called "pillar industry," which includes the electronics industry, to serve as a multiplier for national economic development and to modernize industry structure. Massive incentives are provided for electronics development projects and customs duties have been reduced on all electronics equipment since 1996. As a result, total value of industrial output reached US $37 billion and total electronics export value achieved US $19 billion in 1996. The estimated output value of China's electronics industry will be over US $120 billion by the year 2000 [*China Electronics Industry Yearbook* 1997]. The rapid economic growth, large volume of foreign trade, bold reform measures, and massive infrastructure strategy also point to an enormous market potential and a valuable global partner.

This book documents the technologies, manufacturing procedures, capabilities, and infrastructure that have made China a major player in the Asian electronics industry. This book covers the major segments of the industry: semiconductors, packaging, printed circuit boards, computer hardware and software, telecommunications, and electronic systems. Other topics include the role of government, various associations, research organizations, educational institutions, and major electronics companies. In addition, this study examines the roles that government, associations, research organizations, educational institutions, science and technology information networks, and major companies have played in establishing an infrastructure where the industry can flourish.

This book is intended for readers interested in the historical development, current status, and future growth of China's electronics industry. Engineers, corporate planners, business managers, technologists, and policymakers in the electronics industry will find this book useful in assisting them to:

- comprehend the historical development, current status, and future growth of China's electronics industry
- understand the cultural, economic, and technological factors that drive and inhibit market access and success in China
- make decisions on strategic issues such as market entry, establishing joint ventures or strategic alliances with Chinese electronics companies in order to access the world's largest emerging market
- formulate strategy to cooperate and compete in the global electronics industry

This book features detailed coverage of the important aspects of the Chinese electronics industry and:

- demonstrates how various factors, such as political structures, government policy, science and technology development, education, and labor force, have contributed to the growth and performance of the industry
- reviews Chinese economy in the post-reform period, including general economic status, specialized economic zones, monetary and fiscal policies, foreign direct investment and trade, as well as Sino-U.S. economic relations
- outlines historical development of China's electronics industry, foreign trade and investment in the electronics industry, and national planning on the development and management of China's electronics industry
- evaluates major segments of China's electronics industry, such as semiconductors, computer hardware and software, telecommunications, and electronics systems
- includes valuable site reports for key companies and other organizations
- provides statistical information and numerous tables and figures that illustrate the text.

A brief description of the organization of the book and the topics in the chapters follows.

General information
Chapter 1 provides a brief overview of China's geography, population and major ethnic groups, language, religion, education system, and political structure.

The Chinese economy
Chapter 2 presents the growth of China's economy in the post—reform period, and the use of policy instruments to promote international trade and to encourage foreign direct investment, especially in the electronic industrial sector. Topics include the current economic status, China's foreign trade relations, and the development of special economic zones. Next, the development of China's modern monetary and banking systems, reform of the foreign exchange system, and issues of public finance, such as taxes and import tariffs, are discussed accordingly. The last section summarizes trade friction between the U.S. and China.

Science and technology in China
Chapter 3 outlines China's science and technology infrastructure, current status and goals of national science and technology policy, and the national management of China's electronics industry.

Historical development of China's electronics industry
Chapter 4 reviews the historical development of the Chinese electronics industry that has led to its current status. Topics in this chapter include the roles of China's Five-Year National Development Plans and the foreign trade and investment in the development of China's electronics industry. Major national electronics projects, such as the "Golden Projects" for information management and the "Three Gorges Electronic System Project," are also discussed.

Semiconductors in China
Chapter 5 discusses the status of technological development and evaluates the future growth of the semiconductor industry in China. The emphasis of this chapter is on the government involvement in the development of domestic industry infrastructure and the cooperative relationships between the domestic manufacturers and foreign multinationals in designing, manufacturing, and testing semiconductor components. Electronic packaging and assembly are also discussed in this chapter.

China's emerging information electronics industry
China is seeking to become one of the "Eastern giants" in computers and software. Chapter 6 presents China's emerging information electronics industry, which includes both the computer hardware and software industrial sectors. Topics in this chapter include technology developments in hardware and software, intensive competitions among major domestic manufacturers

and foreign competitors, government-sponsored institutions in assisting the development of China's information electronics technology, and the major problems in China's software industry as well as the key research projects for the future.

The development of China's communications industry
Chapter 7 discusses the development of China's communications industry based on three major industrial segments: telecommunications, data communications, and mobile communications. Discussion topics consist of the periods of national telecommunications development, the use of the Internet and the growing electronic commerce in China, as well as the competition in technological network standards in China's mobile communications market. In addition, foreign joint ventures in China's communication market and the development goals for China's communications industry are also presented.

ACKNOWLEDGMENTS

We are indebted to our schools and professional colleagues for providing us with the time and support for writing this book. Special thanks go to Dr. Linjun Wu of the Institute of International Relations at the National Chengchi University in Taipei, Taiwan, and Dr. Lan Xue of the Tsinghua University in Beijing, China, for providing us valuable information on this subject. We are extremely grateful to the State Science and Technology Commission of the People's Republic of China, the Technological Sciences and Engineering School at the Fudan University, and the Chinese Embassy in Washington D.C. for providing up-to-date information on Chinese economic growth, electronics technology development, and industry strategy and structure. We also extend our sincere appreciation to those who reviewed the draft of this book and offered many suggestions for improvement. We also thank Meerra Ganeshan for helping us in the preparation of the manuscript. Last, but certainly not least, we owe much to our families for their constant encouragement and support.

Contents

Chapter 1

GENERAL INFORMATION

The People's Republic of China, hereafter referred to simply as China, is situated in the eastern part of Asia, on the west coast of the Pacific Ocean. China is the world's most populous and the third largest country next only to Russia and Canada. Most of China is located in the temperate zone. Some parts of south China are located in tropical and subtropical zones, while the northern part is near the frigid zone. Politically, China is under the leadership of the Communist Party and its administration is based on a three-level system dividing the nation into administrative units of three different sizes: provinces, counties, and townships. This chapter provides general information about China's geography, population and ethnic groups, language, religion, education system, political structure, and a brief overview of her modern history.

1.1 GEOGRAPHY

China is centrally located in East Asia. Its frontiers touch most Asian countries: Russia and Mongolia on the north; Kazakhstan, Kyrgyzstan, Tajikistan, and Afghanistan on the west; Pakistan, India, Nepal, and Bhutan in South Asia; Burma, Thailand, Laos, and Vietnam in Southeast Asia; Korea on the east; and Japan, the Philippines, and Indonesia across the East and South China Seas. On the other hand, China is geographically quite isolated. It is bounded by the oceans on the east, impassable gorges on its southern border with Burma, the inhospitable plateau of Tibet to the south and west, and arid and sparsely populated lands to the north and northwest.

In area, China is 3.6 million miles square (9.6 million kilometers square), slightly larger than the United States and only smaller than Russia and Canada. Overall, its terrain slopes from west to east, from the four-mile-high mountains of Tibet, through high plateaus and desert, to hills and plains and finally the deltas of the east coast. The country extends 2,300 miles from north to south; the northern-most province, Helongjiang, lies only 13 degrees from the Arctic Circle, while the southernmost province, Guangdong, is in the tropical zone.

1

Over 40 percent of China is mountainous or hilly. Its most important mountain range is the Qinling, extending east from the great Kun Lun system of north Tibet. True plains are found in north and northeast China, the Yangtze River system, and the Sichuan basin. Like the mountains, most rivers run west to east. The two major rivers are the Yellow River (Huang Ho) in north China and the Yangtze (Chang Jiang) in central China. The Yellow River rises from the Tibet plateau and is unnavigable for most of its 2,700 miles. Its bed has been elevated above the surrounding land by soil carried by the river from north China, and it is liable to flooding despite extensive dikes. The Yangtze River, China's longest river (3,200 miles), is navigable by large ships year-round for a thousand miles from its mouth in the South China Sea. China's mountains, rivers, and other geographical features have divided it into distinct north and south regions, marked by differences in climate, agriculture, culinary traditions, language, politics, and culture.

In terms of natural resources, grasslands cover 400 million hectares, and forests cover about 129 million hectares, or 13.4 percent of China's land. China's mineral reserves rank third in the world, including coal (>1 trillion tons), iron (~50 billion tons), salt (402 billion tons), oil (427 fields), natural gas (125 fields), and nonferrous metals such as tungsten, tin, antimony, zinc, molybdenum, lead, and mercury [Qin 1997]. Inland water from rivers and underground springs comprises 1.82 percent of China's land surface and can provide 6.75 billion kW of energy, of which 3.79 billion kW are developed.

Cultivated lands cover 94.91 million hectares, or about 10 percent of the area of China, mostly in the northeast, north China, and middle-lower Yangtze plains, the Pearl River (Zhujiang) Delta, and the Sichuan Basin [Qin 1997]. Agricultural products are primarily wheat and corn in northern China and rice in southern China, and a large variety of vegetables and fruit in both regions. Pork is the major source of meat protein; minor sources are beef, mutton, and fish. Dry fields constitute about three quarters of the cultivated land in China; wet or paddy fields comprise the remainder. From 1978 to 1994 the area of dry fields diminished ~6.5 percent and the area of paddy fields diminished ~2.5 percent, due mostly to construction in suburban and rural areas.

1.2   POPULATION AND ETHNIC GROUPS

This section presents the profiles of China's population and ethnic groups. The controversial issue of controlling the rate of growth of China's population is also discussed.

1.2.1   Population Profiles

One of China's great resources as well as a source of concern is the sheer size of its population of 1.2 billion (excluding the peoples of Taiwan, Hong Kong, and Macao) — over a fifth of the total population of the world.

Figure 1.1 shows the total population figures in recent decades. Figure 1.2 shows the corresponding natural growth rates (NGR, which is equal to birth minus death rates). China's NGR peaked at 28.38 per 1,000 persons in 1965, then decreased drastically during the 1970s. In 1994 the NGR was 11.21. It is now slightly higher in the countryside, a reversal from the trend up to 1978, when the urban NGR comprised 62 percent of the national rate. Since 1952, the ratio of males to females has averaged 51.45 to 48.54, narrowing since 1952.

The growth of China's urban areas has accelerated dramatically since economic reforms were instituted in the late 1970s and early 1980s. Prior to that time, urban migration was tightly controlled to prevent runaway urban growth (over 80 percent of the population was rural). Urban populations had strict rationing of foodstuffs. In the 1990s, high demand for labor in urban areas has provided ample job opportunities for "illegal" residents, and the vanishing of rationing makes their lives in the cities easier.

Urbanization and attendant declines in arable lands are ongoing trends of concern to both government and populace, especially since China's arable land is already limited and its population large. Environmental preservation is also a concern.

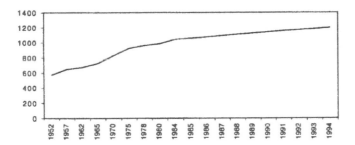

Figure 1.1 Total population (millions) [PRC 1995].

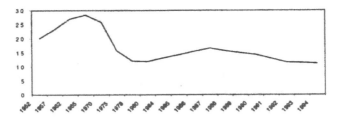

Figure 1.2 Population natural growth rate (per 1000) [PRC 1995].

Table 1.1 shows various population figures of interest in comparison with figures for the United States. Figure 1.3 shows the age distribution of the population in 1990. The majority of the population is in the working category (ages 15-55 for women, 15-60 for men). China has approximately 630 million in its work force: 52 percent work in agriculture and forestry, 23 percent work in industry and 14.6 in services, including commerce [DOS 1997]. The notable changes since 1953 are the decrease in the percentage of the young and an increase in the percentage of elderly. As of 1996, life expectancy in China was about 68.7 for men and 73.1 for women, which is similar to that of the world's developed nations [*Beijing Review*, 1997].

Table 1.1
Living Indicators in 1995

	China	United States
Population	1.2 billion	263.8 million
Annual population growth	1.05%	1.02%
Life expectancy (year)	68.08	75.99
Infant mortality (per 1,000 Birth)	52.1	7.88
Fertility (children per women)	1.84	2.08
Literacy	78%	97%
People per telephone	36.4	1.3
People per television set	6.7	1.2
Urban population	30 %	76%
Unemployment rate	2.9%	5.6%
GDP per capita (U.S $)	$585	$25,900

Source: *Newsweek* 1997

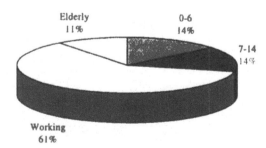

Figure 1.3 Age distribution (1990) — note that elderly is defined as over 55 for women, over 60 for men [PRC 1995].

1.2.2 Ethnic Groups

The Chinese nation is composed of 56 different ethnic groups. The majority of the population is of Han ethnicity, who in 1990 made up about 92 percent of the total Chinese population, down slightly from about 94 percent in 1953. The largest minority group is the Zhuang, which numbers over 15 million and lives mostly in Guangxi Zhuang autonomous region. Twelve other ethnic groups number over one million each, including the Meng (Mongolians), Hui (Muslims), Zang (Tibetans), Uygurs, Miaos, Koreans, and Man (Manchus). Smaller groups include Russians and Hoches. Minority peoples mainly live in western China; a small number live in the north and northeast and on islands off the southeastern coast.

1.2.3 The Issue of Population Size

The Chinese government regards the large size of the country's population as a multidimensional issue. The growth of the population may be the strongest evidence of the success of the Communist regime, of the most prolonged peaceful period China has enjoyed in recent centuries, and the effectiveness of government measures to feed the nation and provide the people with basic health care. The sheer size of the population has provided security to the nation, and during the recent economic reform, it has been exactly the huge population that has created the cheap labor and immense consumer market so attractive to domestic and multinational businesses.

Nevertheless, the government has long recognized the practical need to slow the population's rate of growth. It has restricted families to one child each, except among some minorities and in rural areas, where parents were allowed to have a second child but only many years after the birth of the first. As a result, the population growth rate in 1995 was 1.05 percent [DOS 1997].

1.3 LANGUAGE

To China, with the world's largest national population spread over a vast area and separated by geographic barriers, language standardization is a key to national unity. It is believed that written Chinese was first "unified" more than 2000 years ago, and China has a long and rich written tradition. The process of modernizing Chinese phonology, vocabulary, and grammar began in earnest early in this century. After decades of striving by Mainland China and Taiwan, basic modern standards for written and spoken Chinese have been established, despite some differences in approach between China and Taiwan and certain remaining debates.

1.3.1 Languages of China

There are seven major Chinese language groups (*Putonghua, Wu, Yue, Xiang, Min, Hakka* or *Kejia,* and *Gan*) and hundreds of regional dialects in China — there are said to be 108 dialects in the province of Fujian alone.

Except for *Putonghua,* no other Chinese major language group is spoken by more than about 8.5 percent of the population, and most by less than 5 percent. Although all Chinese dialects are tonal, the number and inflection of tones differ markedly between the various dialects, along with significant differences in pronunciation of words that make most dialects almost unintelligible to speakers of another. For this reason, a standard pronunciation is learned by the majority of Chinese speakers the world over, often in addition to at least one other dialect. The standard pronunciation of Chinese is called Mandarin in the West; *Putonghua,* or common accent, in China; and *Guoyu,* or national language, in Taiwan. *Putonghua* is based on the language spoken in China's capital, and has been China's official language for centuries. *Putonghua* is spoken by at least 71.5 percent of China's population.[1]

Some languages in China are spoken by tiny percentages of the people, including the five major minority languages, Tibetan, Mongolian, Uighur, Zhuang, and Korean. Officially, minorities are encouraged to preserve their own languages. For the first three years in elementary school in minority regions, minority languages are used as the medium of education, and they are taught as a subject through middle school. Newspapers, books, and magazines are published and TV or radio programs are broadcast in minority languages throughout the autonomous regions.

Of the foreign languages spoken in China, English has been the most popular since the early 20th century. In the 1950s, Russian replaced English and was taught in most middle schools and higher education institutes for a decade. With the policy of economic reform and opening up to the outside world, English is ever more popular, especially in major cities. Japanese may be the second most popular foreign language. Other foreign languages such as French, Spanish, and German are only taught at the university level. With the recent surge in foreign imports and international exposure, a flood of foreign words has been imported into the Chinese language that imitate the original sounds but often carry a hint of Chinese meaning.

1.3.2   Language Reform

China's language modernization efforts include a fairly successful attempt to simplify the pictographic characters in an attempt to make learning easier and thus increase literacy. Simplified characters number 2,238, about one-third of the 7-8000 characters required to write modern Chinese. Critics of China's simplification program point to the experience of Taiwan, which simultaneously achieved high literacy and rejected simplification, maintaining that while there are some benefits from the simplified characters, the cultural cost is too high, and also that simplified characters are more easily confused with each other.

---

[1] In the list of those groups, the percentage of population speaking in a dialect is estimated based upon a total of Han Chinese population of 950 million.

Another language modernization effort in China concerns writing directions. In China today, with some specific exceptions, Chinese is written in a direction consistent with that of Romance languages, i.e., in horizontal rows, characters written (and read) from left to right, and rows from the top down. Newspapers, however, are written vertically, top-down, although the columns are read from left to right rather than the traditional right to left.

Among efforts to reform Chinese in this century are failed attempts to romanize the written form, later reduced to the purpose of standardizing pronunciation. Taught in all schools in China today, the sound of each character is represented in romanized written form by *Pinying*.[2] With the ever-spreading use of the computer, some think it is time to finally romanize all written Chinese; however, with computer-voice interfaces already possible, there may soon be no need to key in text character by character. Most Chinese feel that romanizing the written Chinese characters would be impractical given the large number of homonyms in the language, would degrade understanding of both obvious and subtle meanings, and would be an incalculable cultural loss, given the enormous historical, political, and creative value of China's pictographic written language.

1.4   RELIGION

Originally, there was no such concept as religion in China. The word religion did not exist in the Chinese language until modern times, when scholars tried to match a term to the Western concept. Indigenous Chinese "religious" pursuits like practice of Confucianism and Taoism and worship of ancestors, gods, and natural phenomena are not religions in the Western sense but have religious overtones, and are now called Chinese religions simply for convenience. Religions imported from abroad were Buddhism first, followed by Islam and Christianity. The character of expression of Chinese religions is above all a manifestation of the Chinese culture. Except for professional religious practitioners living apart in monasteries, religion in China is very much woven into the broad fabric of family and social life.

1.4.1   Chinese Religions

Chinese religions can be divided in two types: the philosophical type, exemplified by Confucianism and Taoism, and the human type, exemplified by worship of ancestors, emperors, and gods.

Confucianism was "founded" by Confucius (551-479 BCE) during the Zhou dynasty (~1123-221BCE). It has dominated through most of the history of China, strongly influenced East Asian Buddhism, and spread over east and southeast Asia and other parts of the world. Taoism is also believed to have

---

[2] *Pinying* is used in this book to romanize Chinese words. In Taiwan, the romanized spelling is the Wade-Giles method, and another system named *Zhuyinzimu* (phonetic alphabet) is used to define the standard pronunciation.

started in the Zhou dynasty, about the same time Confucianism was established. The first undisputed master was *Laozi*, the "Old Master." Taoism was the dominant thought in the Qin dynasty (221-206 BCE). It flourished over the next several dynasties and has since coexisted with Confucianism and Buddhism.

Ancestor worship can be traced back to the Shang dynasty (~1751-1111 BCE). The central importance of the family is a distinguishing characteristic of Chinese society, and the function of the ancestral cult is certainly one of the distinguishing characteristics of the Chinese family. Ancestor worship as a religious act is more a family practice than individual choice. Put differently, family religion is basic; individual and communal religions are secondary.

Regarding emperor worship and worship of deities, it was believed by most Chinese that the emperor, in addition to his official duties, had an essential role to play in mediating between the forces of nature and the lives of people. Other people such as famous statesmen, heroic generals, just and merciful magistrates, patriotic scholars, or even legendary figures may become gods after their deaths. Temples have been built for them to provide places for people to worship.

Cutting across "religions," social groups, and voluntary institutions is a series of annual festivals observed by the entire nation, particularly the following five. These may have originally derived, at least partly, from religious origins. Their dates are set by the lunar calendar. New Year's is by far the most important and elaborate of all Chinese festivals, concluding with the first full moon in the new year, celebrated as the Lantern Festival. The Qing Ming festival at the beginning of the spring season is the most important time to visit ancestral tombs, clean and renovate them, and sacrifice (foods, etc.) to them. The Double Five festival, also known as the Dragon Boat Festival, has long been understood as a re-enactment of the search for the drowned poet Qu Yuan, who committed suicide because his honest counsel was spurned by his lord. The Moon festival is celebrated in mid-Autumn, when the moon is supposed to be at its fullest and clearest.

1.4.2   Foreign Religions

Buddhism was imported from India probably in the Later Han dynasty (23 BCE - 220 CE). It flourished over several dynasties and reached the zenith of its influence during the Tang dynasty (618-907). It has been much indigenized and is the most accepted foreign religion, both officially and by the populace. Old temples have been restored and renovated and new ones built. Some estimate that there are more than 100 million Buddhists in China; however, given the nonexclusive nature of Buddhism, this figure may be misleading. The practice of Buddhism by ordinary people is mixed with practice of Confucian doctrines and superstitious beliefs.

Two to three percent of the population are Muslim (*Hui*, in Chinese) and many do practice their religious beliefs. Most Muslims reside in the autonomous regions, but they live throughout all parts of China. It is reported that they enjoy the greatest freedom of worship in China, perhaps because they are treated as one of the largest minority groups, and thus Islam in China is identified with a specific ethnic minority.

About 1 percent of China's population is Christian. The official number for 1994 was in the neighborhood of 10 million, more than half of them being Catholics. The unofficial number was estimated at 30 million. The number of Christians has been increasing since 1985.

1.4.3   Official Policies Towards Religion in China

Officially, China claims to be an atheist state with a belief in "historical materialism" (a Marxist term). While freedom of religion was written into China's constitution, religion and freedom have different connotations than the Western definitions. Authorities interpret religious practice as the practice of personal devotion and spiritual self-development only. A keystone of China's current religious policy and a necessary condition for the continued existence of religious organizations is the "Three-Self" concept, meaning religions must be self-governing, self-supporting, and self-propagating. In other words, there cannot be any foreign influences in terms of ideas, policies, or funds.

Superstitious activities are strictly prohibited in China; nevertheless, they also have been increasingly practiced and tolerated in recent years. Under a much more tolerant policy in the past two decades, all religions have been flourishing in China since economic reform.

1.5   EDUCATION

Chinese leaders face three challenges in designing education policy. First, as a developing country, China faces resource allocation tradeoffs between providing large numbers of people with a relatively low standard of education, and training a much smaller number of people in areas that could enhance the nation's economy, political power, and military strength. Second, as an ancient country, China faces curriculum tradeoffs between teaching traditional norms to preserve the culture and national identity, and teaching modern technology and thought to be internationally competitive. Third, the leadership bears the mandate to shape and support the people's belief in the communist party's ideology, policies, and rules, and the directions the Party has chosen for the country in various periods of time (see Dreyer 1996). Many of the complex issues of ideology in education have been displaced by a new spirit of pragmatism and determination to pursue policies conducive to rapid national development. On the other hand, the difficult problems involved in modernizing China's education system and upgrading the literacy and skills of its people in an efficient but equitable way still engender considerable political debate.

1.5.1   Literacy

The literacy rate in 1949, when the communist party came to power in China, was about 20-25 percent. It reportedly doubled by 1955, reached 61.9 percent in 1964 and 76.5 percent in 1982, but declined to 73.3 percent in 1988. In 1998 it was approximately 78 percent. This achievement is viewed by the leadership as being due to the joint efforts of all forces involved, starting from the language reform efforts that began early in the century.

Although the climbing literacy rate is a laudable achievement, the statistics must be interpreted with caution. Literacy is defined as having a minimum of 4.5 years of education, a rather low requirement, especially for teaching a language as difficult as Chinese. Second, the literacy rate in cities is very high; therefore, the literacy rate is much lower in rural areas than the national average. Gender also makes a significant difference. It is reported that 70 percent of illiterates are women living in rural areas.

1.5.2   The Formal Education System

China's formal education system is very similar to that of the West, a result of China having imported the system from the West early this century. Children go to primary school at about 6 to 7 years of age and attend for six years. Attendance at this level is 98.8 percent of primary school age children [Beijing Review 1997]. Primary school is followed by three years of junior middle school and three years of senior middle school, but availability and attendance of the higher level classes starts to drop. Nine years of education are compulsory, but enrollment in junior secondary schools is only 82.4 percent [Beijing Review 1997]. In addition to the above levels, preschool education has been widely available in cities, to the benefit not only of the children but also of the many women in the labor force.

Higher education ranges from two to six years, including both universities and a variety of vocational and technical schools. As of the 1990 census, only 1.4 percent of the population had received university education. According to 1995 data, 1.5 million students annually took a highly competitive exam for one of 500,000 university admissions places [IEEE 1995]. Other data indicate that university admissions in 1996 numbered 966,000, up 40,000 from the year before [Beijing Review 1997]. University classrooms are still too few for the country's educational needs, so many students enroll in a short study or two-year program, or take TV courses. Table 1.2 compares the numbers of schools and enrollments (per 10,000 persons) at different types of schools in 1962 and 1994.

Graduate students numbered 162,000 in 1996, up 17,000 from the year before [Beijing Review 1997]. Because of the shortage of graduate study programs, the government supports many talented graduate students to continue their study abroad to bring new technology into China; the risk is that some students choose not to return.

Table 1.2
Education in China 1962-1994

	1962		1994	
	Number of Schools	Number of Students*	Number of Schools	Number of Students*
Universities	610	83	1,080	279.9
Specialized Secondary Schools	1,514	53.5	3,987	319.8
Regular Secondary Schools	19,521	752.8	82,358	4,981.7
Vocational Schools	3,715	26.7	10,217	405.6
Primary Schools	6,668,318	6,923.9	682,588	12,822.6
Kindergartens	17,564	144.6	174,657	2,630.3
Schools for Blind, Deaf, and Deaf-Mutes	261	1.8	871	7.2

*Per 10,000 persons; Source: China Statistical Yearbook.

1.5.3   Education Expenditures

The percentage of gross national product (GNP) devoted to education in China was only 2.7 percent in 1993. According to United Nations data, the average education expenditure is 6.1 percent of GNP in the developed world, 4.1 percent in developing countries, 4.6 percent in Asia, and 5.7 percent worldwide. In other words, the 1993 education investment in China was not only lower than the international average, but also below that of developing countries. The 1993 Chinese government expenditure on education marked a 58 percent increase in education spending since 1990; however, 85 percent went for administrative costs. Only 3.9 percent went for capital construction.

Schools at all levels, including institutions of higher learning, have been tuition-free up until recent years. In the past several years, various fees have been charged to raise money at all levels. Colleges and universities started charging tuition in 1997. Private schools have also emerged.

1.5.4   Prognosis

While the educational levels of China's population as a whole are rising, as are numbers of persons with undergraduate and postgraduate degrees, economic reform has in some ways slowed and even reversed some of China's education gains. In rural areas, the elementary school dropout rate has increased. Being poorly paid, many intellectuals have left educational institutions for better paying jobs. Many of the brightest youths are seeking opportunities abroad. Disparities between the educational levels of rural and urban dwellers, and of intellectuals and workers are not only an economic but also an ideological and political problem in China, and thus are not likely to be ignored.

The country's low educational mean and its educational disparities are not expected to change rapidly. On the other hand, optimists put their faith in the Chinese culture's emphasis on family and education. Most Chinese families make children's education one of their highest priorities. Also in China's new, more market-oriented economy, the demand for educated and skilled workers at all levels is increasing, and market forces may help channel more resources into the education arena.

1.6   POLITICAL STRUCTURE

China is a communist state whose political institutions are derived from the Soviet Union, the only available model when the Chinese Communist Party came to power. Still, China's political system has many embedded traditional (feudalistic) elements that make it different from the Soviet model, and many changes have taken place. Since the reform movement started and China began opening up to the outside world, the leadership finds it needs more consistent policies in order to compete more efficiently both domestically and internationally. New offices are being established and old ones removed or restructured, and many new laws and regulations are being developed.[3]

1.6.1   The Chinese Communist Party

The Chinese Communist Party (CCP) is granted absolute power by China's constitution, and it is involved throughout the decision-making processes that determine the social, economic, and political goals for the country. It has the exclusive right to legitimize and control all other political organizations.

A conspicuous characteristic of the CCP is its hierarchical, pyramidal, and centralist structure. Formally, there are four organizational levels. At the top of the structure are the party Chairman, the Standing Committee, the Politburo, the Central Committee, and the National Party Congress. At the second level are the provincial and autonomous regional organizations with their committees and congresses. At the third level are the county and city organizations. At the base of the structure are the primary party branches, which cover every cell of the society, such as schools, factories, and communities.

The party's oversight over the government bureaucracy is tight, centralized, and pervasive. More specifically, the party retains the authority to select government officials, propose policy initiatives, and veto policies that emerge from the government.

---

[3] A wide body of literature describes the Chinese political system, for example, Shirk (1993), Dreyer (1996) and Wang (1980).

1.6.2 Administration

China's civil administration consists of three levels presided over by the State Council: (1) the provinces and autonomous regions with their pertinent cities, and municipalities directly under the central government; (2) prefectures with their pertinent cities and counties with their pertinent cities; and (3) townships, national minority townships, and/or towns. Major cities are further divided into districts. China is currently divided into 23 provinces, 5 autonomous regions, and 4 municipalities directly under the Central Government. The capital is Beijing; other major cities are Shanghai, Tianjin, Guanzhou, Shenyang, Wuhan, Chengdu and Chongquing.

The Chinese government bureaucracy makes decisions by a system of "delegation by consensus." Party leaders delegate authority to subordinate government agencies to work out specific policies. If everyone agrees with a proposed policy, the leaders simply ratify it; if some agencies disagree, then it is sent to higher levels for resolution or is tabled. Hierarchical control gives officials an incentive to compromise rather than exercise their veto, especially when the political climate clearly identifies certain preferences.

Figure 1.4 illustrates the hierarchy of China's state administrative bodies.

1.6.2.1 The People's Congress

The National People's Congress (NPC) is theoretically the highest organ of state power in China, although the NPC and local people's congresses (legislative bodies) in general have not exercised much political authority.

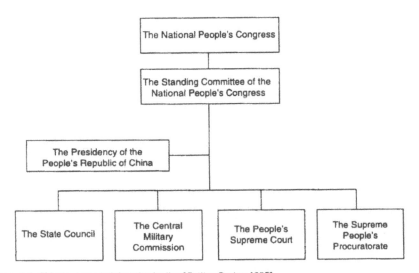

Figure 1.4 China's state administrative bodies [*Beijing Review* 1997].

The NPC exercises legislative power, makes decisions on important national issues, elects the President and Vice President, Premier, and other component members of the State Council, and also elects the Chairman of the Central Military Commission and other members. The local people's congresses at different levels are local administrative organs of state power working under the State Council in provinces, special autonomous areas, counties, and townships. They have the autonomy to decide on important local affairs in their respective administrative areas and to formulate local laws and regulations [Qin 1997].

1.6.2.2 The State Council

The State Council is the nation's highest executive organ, which carries out the will of the party and the NPC and administers the government through functional ministries, commissions, and special agents. It consists of a premier, several vice premiers, ministers and commissioners, a secretary-general, and several deputy secretaries-general.

The State Council implements the laws and resolutions adopted by the NPC and its Standing Committee and reports to them on its work. The President's mandate is to pursue the decisions of the NPC, promulgate statutes, appoint and remove members of State Council, and ratify and abrogate treaties and important agreements reached with foreign states.

Theoretically, the State Council is under the People's Congress; in reality, it carries the party's decisions. An overwhelming majority of important positions are held by party members, with a few exceptions. Nevertheless, it is less a political organ than an administrative one, and concerns itself with economic, diplomatic, and other practical issues.

1.6.2.3 Local Governments

Since the beginning of economic reform, one big change in the political landscape is the weakened central government and strengthened local governments. This is particularly the case in those provinces along the coast that are benefiting the most from reform.

1.6.2.4 Military Organization

The Chinese military consists of the People's Liberation Army (PLA), the People's Armed Policy Force, and the People's Militia. Oversight is exercised by the Central Military Commission. The Central Military Commission has been supervising the work of modernizing and streamlining China's 3-million-strong armed forces. The political role of the military in the People's Republic of China has been substantial, although subordinate to the Party. There has been a traditional lack of clear separation between military and civilian leadership. In addition, the military has always served the country's economic purposes, tracing back to the guerilla period and continuing to the present economic reform period. The highly respected PLA has long been promoted to the whole society as a moral model.

1.6.2.5 Judicial Organization

The state judicial organs are the Supreme People's Court, with its subordinate local people's courts and special people's courts. The state legal supervisory organs are the Supreme People's Procuratorate, with its subordinate regional and special procuratorates, which are responsible for legal supervision of the judicial and penal systems.

1.6.3 Political Reform

Political reform in China has gone through several stages. It appears that some improvements are being made in decision-making processes, making them more fair and transparent, to strengthen the legitimacy of Party rule. Grassroots free elections, already in existence, seem to be spreading over more areas and sectors. There are some hopeful signs that the people may gain more freedom in artistic, journalistic, and other forms of expression. However, freedom of speech and of the media continue to be carefully circumscribed, especially in times of political tension.

1.7 BRIEF OVERVIEW OF MODERN CHINESE HISTORY

This section presents a brief overview of the modern history of China since the mid-19th century. Communist rule under party chairman Mao Zedong and the economic reform after Mao are reviewed in section 1.7.2 and 1.7.3, respectively. Issues regarding China's internal stability and international relations are also discussed in brief at the end.

1.7.1 A Culture in Crisis

For several millennia, China was isolated but self-sufficient. Invaders occasionally occupied China but were, eventually, in some form, assimilated. Dynasty after dynasty, under the dominance of Confucianism and Buddhism, essentials of Chinese culture were reflected in the political structure, social order, moral values, and other traditions that together constituted one of the world's most enduring, homogeneous, and extraordinarily stable civilizations.

The Opium War, begun in 1840, marks the start of modern Chinese history, the salient feature of which has been China's inability to separate itself from external forces, particularly those originating in the West. The confrontation between the Chinese tradition and the West covers almost all aspects of the society — military, moral, cultural, economic, technological, and so forth. Starting in 1840 China for the first time faced a challenge from the West that forced it, through force of superior arms, to open its market and provide resources to the West. Chinese civilization was severely humiliated. Stability could no longer exist. Emperor Guang Xu tried to institute reforms, but failed. Li Hongzhang initiated learning from Western technology, but this was insufficient to defend the country against the West.

Finally, there was the internal dichotomy started at the beginning of this century, which progressed to armed conflict even through the Japanese occupation. On the one hand, Sun Yat-sen and his followers turned to mainstream Western ideology, founded the nationalist party, and in 1911 established the Republic of China in a revolution that ended the Qing dynasty and 2000 years of monarchy in China. On the other hand, Mao Zedong and his followers turned to communism, championing the dispossessed. He led a successful Marxist revolution, similar in many ways to a traditional peasant rebellion, and established the People's Republic of China on October 1, 1949. The Nationalist government and its adherents fled to the island of Taiwan.

1.7.2   China Under Mao Zedong

The Communists under the leadership of their populist leader Mao Zedong moved quickly to integrate all of Chinese society into a People's Republic. The new government unified the nation and in its early years achieved a stability that China had not experienced for generations. With the assistance of the Soviet Union, the People's Republic developed a centrally planned economy in the 1950s that eventually proved inefficient. For a variety of reasons, Soviet leaders severed their ties with Mao in 1962. Soviet suspension of aid was a terrible blow to the Chinese scheme for developing industrial and nuclear technology. As hostility grew between China and the Soviet Union, China strategically turned back to the West in the early 1970s.

Although the author of many admirable achievements, Mao's legacy includes two potent campaigns he set in motion that ultimately brought chaos and severe economic distress to his people: the "Great Leap Forward" to expedite self-reliant economic growth, and the "Cultural Revolution" to carry the communist revolution and class struggle into all aspects of the very traditional Chinese society. These ambitious, idealistic campaigns brought such severe economic, political, and cultural disruption that tens of millions died of starvation, numberless cultural treasures were destroyed, the people's resilience was severely strained, and China fell even further behind other nations in scientific, technological, and economic capability.

1.7.3   Economic Reform After Mao

After Mao Zedong died in 1976, an early protégé but often outcast pragmatic leader, Deng Xiaoping, eventually established his power in 1978. Almost immediately, he began China's transition from central planning to a market-oriented economy, coupled with a new policy of opening to the world. Deng focused the country's energies on "the four modernizations" of agriculture, industry, national defense, and science and technology, in preference to class struggle. The initial focus was on revitalizing agricultural production, followed a few years later by a corresponding program to revitalize industry. This time, the country's direction was chosen voluntarily by the majority of the people, and the overall results have been widely considered a success.

Deng dissolved Mao's famed agricultural communes and leased the land back to private households, although without really privatizing the agriculture market [*Economist* 1997]. The farmers became responsible for output and they had the right to the income produced, but they were not able to buy and sell their land. As a direct result of agricultural reforms, in the following 6-8 years, grain output grew by a third, cotton production almost trebled, fruit production went up by half, and real farm income almost trebled [Rohwer and Rohwer 1996].

In his ensuing industry reforms, Deng encouraged private enterprise, worked to upgrade inefficient management styles and antiquated technologies, implemented price reform, encouraged foreign investments and trade, and even approved a stock market [Rohwer and Rohwer 1996]. However, Deng's early industrial reforms never had the success of the early farming reforms. The industries were owned and controlled almost entirely by the state, and their capital assets could not simply be subdivided as plots of lands. More radical changes would have called into question the role of the Communist party. Even so, Deng launched reforms against the state-owned industries, adopting the policy to attract foreign business, and allowing non-state industries [Rohwer and Rohwer 1996].

Since 1987 China has pursued a high-tech research plan to keep abreast of the world's latest scientific achievements in biological engineering, energy, new materials, information, storage and retrieval, laser technology, astronautics, and automation. The Dengist reforms of 1979-1994 brought an extraordinary rise in living standards and economic growth of 9 percent a year.

1.7.3.1 Ideology vs. Pragmatism

Deng made his reforms not following a preconceived plan or philosophy, but by "crossing the river by feeling the stones," formulating only in retrospect an ideological paradigm for the changes. He rejected the insinuations of communist critics that he was making China capitalist. He insisted that his modernization program was a socialist one designed to make China strong, not democratic [Schell 1997].

With rising expectations among the people due to the economic successes of Deng's programs, the challenges to present leaders, headed by President Jiang Zemin and Premier Zhu Rongji, are to consolidate China's gains and move forward, and to do so in a way that maintains stability. Ongoing reforms are not without anguish and cost — both political and monetary. Efforts at fundamental transformation of economic, governmental, and political organizations cause discontent among some elements of the society and are resisted by those who cling to the "iron rice bowl" of guaranteed lifetime job tenure. Beijing's reform leaders have made repeated calls for party members and government bureaucrats to reform their "ossified thinking" and to adopt modern methods. A younger and more technically oriented generation of managers and leaders is taking over, but this poses a political dilemma and an ongoing debate between those who emphasize ideological correctness and those who stress the need for technical competence. At present, the technocrats are dominant, but their detractors remain. Furthermore, even among the reformers, there are debates concerning the scope, pace, and details of proposed changes. Political confrontation over the reforms has been pervasive and, to many foreign observers, the issues are confusing.

One of the key issues surrounding reform is that achieving the country's ambitious economic goals requires China to reject its concept of "self-reliance," and with it, historical definitions of socialist economic organization. Foreign specialists have been invited to assist in the modernization process. The country has entered into the milieu of international bank loans, joint ventures, and a whole panoply of once-abhorred capitalist economic practices that, in effect, undermine the government's communist legitimacy.

Reformers are therefore still struggling with Deng Xiaoping's dilemma of needing to establish a sound ideological basis for implementing needed economic reforms that do not appear consistent with communist principles. Although Deng himself worked out a complex theory that placed the reforms on a continuum to attainment of true socialism, the most enduring arguments appear to be variations of the proposition that Chinese values can and should be applied to Western technology-based systems. This philosophy is reminiscent of the traditional *tiyong* (substance versus form) formula evoked in the late 19th century reform period.

In addition to the complex political issues involved in applying economic reforms, it is undeniable that developing and successfully applying technological expertise costs vast sums of money and requires hard work and sacrifice on the part of the Chinese people. In order to maintain popular support for their programs, the reformists must produce tangible results in terms of raising incomes and standards of living. In essence, the central problem for the reformists is on the one hand to create unity and support for the scope and pace of the reform program among party members, while on the other hand delivering material results to the broad masses of people amid economic experimentation and mounting inflation. Failure to achieve this balance and to make successful mid-course corrections could prove disastrous for the reform leadership.

1.7.3.2 Evaluation of the Reforms

So far, it seems that China has been quite successful in maintaining this balance. Self-proclaimed successes of the reformers in the 1980s include improvements in both rural and urban life, adjustment of the structures of ownership, diversification of methods of operation, and introduction of more people into the decision-making process. As market mechanisms became an important part of the newly reformed planning system, products have circulated more freely and the commodity market has rapidly improved.

The government has made inroads on rationalizing prices, revamping the wage structure, and reforming the financial and taxation systems. The policy of opening up to the outside world has brought a significant expansion of economic, technological, and trade relations with other countries. Reforms of scientific, technological, and educational institutions have rounded out the successes of the Deng-inspired reforms. Also, for the first time in modern Chinese history, reforms are being placed on the firm basis of a rational body of law and a carefully codified judicial system. Although reform and liberalization have left the once more strictly regimented society open to abuses, the new system of laws and judicial organizations continue to foster the relatively stable domestic environment and favorable investment climate that China needs in order to realize its modernization goals.

Amid these successes, authorities have admitted that there have been difficulties and failures in attempting simultaneously to change the basic economic structure and to avoid the disruptions and declines in production that marked previous ill-conceived "leftist experiments." China's size and increasing economic development has rendered central economic planning ineffective, and the absence of markets and a modern banking system have left the central authorities few tools with which to manage the economy. A realistic pricing system that accurately reflects levels of supply and demand and the value of scarce resources is yet to be implemented. Tremendous pent-up demand for consumer goods and a lack of effective controls on investment and capital grants to local factories have unleashed inflationary pressures that the government finds difficult to contain. Efforts to transform lethargic state factories into efficient enterprises responsible for their own profits and losses are hampered by shortages of qualified managers and by the lack of both a legal framework for contracts and a consistent and predictable taxation system.

On a far more discordant note, the state has felt obliged to repress a growing democratization drive among its people, which was set in motion by the success of the economic reforms as well as by increasing contact with the outside world. Early into the period of economic reform, China's students began to call for freer artistic and literary expression and for more democratic processes, and even openly criticized the party. This in turn led to the Tiananmen Square incident on June 4, 1989.

1.7.4   The Issue of Stability

Of all their various concerns, most people in China desire stability. The communist party seeks stability to safeguard its power; the people want stability to secure their gains from reform and to be able to pursue their lives in peace; foreign investors and businesspeople require stability to protect their investments and businesses in China.

A number of theories about threats to China's stability are considered unrealistic by most Chinese. These include worries about power struggles within the communist party; about south China seeking separation from the northern central government; about re-unification with Taiwan; and about China's border and territorial disputes with India, Russia, Japan, and Vietnam, or others. The problems in these scenarios are not now and are not likely to become acute. Other destabilizing factors are real concerns, however. Reform and rationalization of unprofitable state-owned enterprises and traditional agricultural practices threatens massive unemployment, especially in the Northeast where those enterprises are concentrated, and among rural unskilled laborers. Corruption within the party can undermine its authority. Issues concerning Tibet and minorities in the Northwest are worrisome to the central government as well as to domestic and foreign interested parties. China's government is cognizant of these complex problems and is working to solve them in ways that avoid turmoil.

1.7.5   Hong Kong, Macao, and Taiwan
In July 1997, Great Britain handed Hong Kong back to China. Under the principle of "one country, two systems," Hong Kong will remain as a Special Administrative Region (SAR) of China. Since Hong Kong has one of Asia's highest standards of living and for years has based its economic success on international trade, the return of Hong Kong adds vitality to China's economy and a valuable source of communication with the outside world. Already, Hong Kong handles 40-50 percent of China's foreign trade, and it is China's most valuable port. Some think China's "one country, two systems" economic arrangements with Hong Kong may represent a model for reunification with Taiwan.

Hong Kong is also the stepping stone between the economies of Taiwan and China. In 1995 Hong Kong was Taiwan's third largest trading partner and second largest export market (21.7 percent in 1994) [Hong Kong Trade Development Council 1997]. World Bank 1995 statistics showed that Taiwan ranked third in market share (11 percent) in mainland China, trailing only Japan (22 percent) and the United States (12 percent). Over half of all container shipments from Taiwan pass through Hong Kong for destinations in China or other parts of the world. As much as 70 percent of Taiwan's exports to Hong Kong are shipped onward into the mainland, and many of the 1.9 million Taiwanese visitors to Hong Kong are bound for the mainland. This indirect commerce between China and Taiwan via Hong Kong is a significant factor in the economies of all three entities.

China has been explicit in its intentions concerning eventual reunification with Taiwan, a region that is now one of the richest in Asia. At the 15th National People's Congress of the Communist Party of China (CPC) in September 1997, China's leadership reiterated that holding steadfast to the principle of the "one-China" policy remains the key to improving cross-straits relationships. Peaceful reunification of Taiwan with Mainland China and "one country, two systems" policies toward Taiwan, a Party spokesman said, are consistent with China's government, and the course would not change.

The Taiwanese, in general, think otherwise. Polls quoted by Taiwanese newspapers report the majority of adults favor maintaining Taiwan's political status quo for the time being. While some prefer independence either immediately or in the near future, others prefer unification with China when the economic and political structures of Mainland China and Taiwan are more in tune with each other. Many Taiwanese want to live under the existing conditions in Taiwan and reject steps toward unification until a free and democratic system is implemented in the mainland.

The most accurate way to describe the Taiwan-Mainland China situation may be that it is fraught with both danger and opportunity. Reunification represents a dream for the nation of China to be truly strong in the world arena. Some can imagine a two-party political system born of a theoretical reunification; others may imagine leaving their names in Chinese history by accomplishing reunification. The facts are not so promising. Historical efforts at cooperation between the Communist and the Nationalist parties have failed, ending with civil war. The ideological differences between the two parties are fundamental, and distrust is deep. Current differences in economic development and political structure only make unification more difficult. Reunification by military means is unlikely but remains a continuing concern to Taiwan and its U.S. supporters.

1.7.6   Other International Relations Issues

National security has been a key determinant of Chinese planning since 1949 and national defense is part of the Four Modernizations. In the 1970s China moved to develop intercontinental ballistic missiles, nuclear submarines, and other strategic forces; and to acquire sophisticated foreign technologies with military applications. On the diplomatic front, China in the 1980s used improved bilateral relations and a variety of international forums to project its "independent foreign policy of peace" while opening up to the outside world. China's increasing involvement in international affairs has gone hand in hand with the world's increasing involvement in China's economic development. A goal of China's leaders is to carefully manage international involvement in its economy while simultaneously maintaining its independence and increasing its stature in the international community.

Chapter 2

CHINA'S ECONOMY IN THE POST REFORM PERIOD

China has operated under a planned economy since the 1950s. Eight "five-year national economic and social development plans" have been implemented; the ninth five-year plan runs from 1996 through 2000. Nevertheless, true economic reform began in 1979 under the leadership of Deng Xiaoping. The goals of national economic policy have been to establish a more or less ideal "socialist market economy" and promote national economic growth that is sustained, rapid, and sound. The Chinese Communist Party's 14th National Congress in 1992 made three policy decisions of far-reaching significance: (1) seize opportunities to speed up economic development; (2) define establishment of a socialist market economy as the goal of economic restructuring in China; and (3) establish within the whole Party the dominance of Deng Xiaoping's theory of building socialism with Chinese characteristics.

The 9th five-year plan (1996-2000) and "Outline of Long-Term Targets for National Development by 2010," which were formulated by the National People's Congress in March 1996, are the first middle- and long-term plans to be enforced as China develops a market economy. These two national plans have further defined industrial policy in terms of five "pillar industries" — machinery, electronics, petroleum and chemicals, automobiles, and construction. These are the key building blocks of China's industrial policy for the 1990s and beyond.

Overall, the Chinese economy has been In a state of transition away from the central planning model for twenty years now. The central government has much less control over the national economy through its various policy instruments such as budgetary control. The power of local governments has been rising in making their own economic policy. The central government has not yet fully mastered managing a market economy; nevertheless, the role of centrally planned industrial policy continues to be diminished by broader market forces.

2.1 GENERAL ECONOMIC CONDITIONS

China's economy has grown at an average rate of nine percent per year since the 1979 economic reforms, with growth rates of more than 10 percent from 1992-95 (see Table 2.1). China's gross domestic products (GDP) in

1997 grew by 8.8 percent, slower than the 10 percent growth projected at the beginning of the year by government and private economists. Overall, however, China's growth in the 1990s has been faster than that of any other economy in the world.

China has achieved a "soft landing" of single-digit inflation and stable growth in 1996 and 1997. Table 2.1 shows that the retail price index, which exceeded 21 percent in 1994, rose just 0.8 percent in 1997. The consumer price index has decreased from more than 24 percent in 1994 to less than 3 percent in 1997. Per capita income has risen significantly since the 1980s, and the official unemployment rate stood at 4 percent in 1997.

Although several Asian economies have experienced fiscal problems in 1997-98, China's economy and currency value have stood firm. Currency value is controlled by the government, so a catastrophic decline is unlikely. Also, most investments in China are in plants and equipment, not short-term bonds, so a shortage of hard currency to pay foreign debts is improbable.

A major economic problem facing China are inefficiencies in its state-owned enterprises (SOEs), which range from large manufacturing plants to small service providers, employ 100 million workers, and account for about 30 percent of China's GDP. Many operate in the red and rely on government subsidies to stay in business. The bad debt accumulated by SOEs threatens to destabilize China's entire banking system. In addition, subsidies to SOEs constitute the largest barrier to China achieving its major international economic objective, joining the World Trade Organization (WTO), because the subsidies are beyond maximum levels allowed by the WTO. Therefore, China must solve the SOE problem in order to maintain economic stability and pave the way to full membership in the WTO and the global economy.

2.2 FOREIGN ECONOMIC RELATIONS AND TRADE

Following the successful strategies of East Asian newly industrialized economies, China has implemented an export-led growth strategy. The increasing openness of the Chinese economy, measured by the trade per GDP ratio, has been an important factor in its growth.

2.2.1 Current Trading Status

Since 1979, China has aggressively expanded its foreign trade and technical cooperation with other countries, as reflected in the trade statistics shown in Table 2.2. In 1996 China's foreign trade totaled $289.9 billion, an increase of 3.2 percent over 1995. Total import and export volume for 1997 was estimated to exceed $325 billion, a rise of seven percent over that of 1996. In terms of international trade volume, China ranked 10th in the world on the 1997 world trade list issued by the WTO, up from 32nd in 1978.

Table 2.1
General Economic Indicators of China (US $ billions)

Year	GDP*	Real GDP Growth (%)	Retail Price Index	Consumer Price Index	Urban per Capita Income*	Rural per Capita Income*	Unemploy-ment Rate (%)
1991	406.3	9.3	2.9	3.4	290.3	133.2	2.3
1992	482.6	14.2	5.4	6.4	330.8	142.0	2.3
1993	601.3	13.5	13.2	14.7	405.6	160.0	2.6
1994	542.5	11.8	21.7	24.1	368.8	141.6	2.8
1995	700.3	10.2	14.8	17.1	466.2	188.9	2.9
1996	826.4	9.7	6.1	8.3	527.3	232.0	3.0
1997	904.1	8.8	0.8	2.8	621.5	251.5	4.0

Note: Based on China's official exchange rates: 1991 (5.32), 1992 (5.52), 1993 (5.76), 1994 (8.62), 1995 (8.35), 1996 (8.30), and 1997 (8.27).
Source: China State Statistical Bureau (SSB), 1998

The composition of goods traded has changed since the late 1970s. The proportion of industrial finished products among all goods exported rose from 46.5 percent in 1978 to 79.9 percent in 1992 and 85.5 percent in 1996; at the same time, the proportion of primary products dropped from 53.5 percent in 1978 to 20.1 percent in 1992 and 14.5 percent in 1996.

Overall, the share of mechanical and electronic products among exports rose from 29.5 percent in 1995 to 31.9 percent in 1996, continuing to be the largest category of export commodities. In terms of imports, industrial products have consistently accounted for about 80 percent of goods imported. The dominant imports are raw materials, energy products, and transportation equipment, which are in short supply in the growing domestic market.

As of 1997, China has established trade relationships with more than 220 countries and regions around the world. Japan is China's biggest trading partner; trade between the two countries in 1996 exceeded $60 billion, about 20.7 percent of China's trade, which was a 4.5 percent increase over the previous year. China's other major trade partners are the United States, the European Community, the Republic of Korea, Taiwan, Singapore, and Russia.

Although China's foreign trade surplus has continued to grow since 1990 and reached $20 billion in 1997, that surplus is expected to decrease as a result of actions to slash tariffs and reduce interest rates. On October 1, 1997, China's average tariff level was reduced from 23 to 17 percent. These measures were designed to stimulate domestic consumption and investment and bolster the demand for more imports. Another major incentive for more imports was the resumption in 1998 of free duty on imports of capital goods for high-tech projects. The increased imports are primarily earmarked for ongoing development of China's high-tech industries and technology improvements in her state-owned enterprises. A third inducement to increase imports is China's bid for admission into the WTO.

Table 2.2
Foreign Trade Volume in China ($ Billions)

Year	Import	Export	Total Trade Volume	Trade/GNP (%)	Foreign Exchange Reserve
1950	0.58	0.55	1.13	N/A	N/A
1965	2.02	2.23	4.25	N/A	N/A
1978	10.89	9.75	20.64	9.07	N/A
1980	20.02	18.12	38.14	13.05	N/A
1985	42.25	27.35	69.60	23.88	2.66
1990	53.35	62.09	115.44	31.20	11.10
1991	63.79	71.84	135.63	35.68	21.71
1992	80.61	85.00	165.61	38.09	19.44
1993	103.96	91.74	195.70	40.10	21.11
1994	115.69	121.04	236.73	45.42	51.62
1995	132.08	148.77	280.85	41.02	73.60
1996	138.83	151.07	289.90	35.49	105.03
1997*	142.00	183.00	325.00	36.00	140.00

Source: China State Statistical Bureau, 1998.   * Estimated
Note: The increase in trade/GNP ratio from 1978-1994 partly reflects the depreciation of
Chinese currency (yuan).   The decline of this ratio since 1994 mostly reflects the
appreciation of yuan.

2.2.2 Trade Policy and Administration

Before 1979, government planners adhered to the principle that too
much involvement in foreign trade could harm China's centrally planned
economy through external fluctuations and dependence on foreign markets.
Therefore, trade policy was biased in favor of import substitution and against
export promotion. However, Chinese leaders have gradually recognized that
in order to modernize its economy and transform it into an industrial country,
China must actively participate in world trade. The new trade policy has
achieved extraordinary success in terms of economic growth and per capita
income. China has been able to greatly expand domestic production and
acquisition of foreign technology and know-how.

Nevertheless, maintaining economic independence is still a concern.
Foreign companies are not permitted to directly engage in trade in China,
other than marketing goods they have manufactured in China. Foreign
exporters must use a domestic agent for both importation into China and
marketing within China, or handle their own sales through a representative
office. Only companies authorized by the central government to handle
export and import business may sign import and export contracts.

Before 1979 China had a highly centralized trade management system.
Foreign trade was monopolized by a dozen state-owned foreign trading
corporations (FTCs), each responsible for carrying out trade in specified
areas at the national level. That system has been gradually decentralized
since the inception of the open-door policy. In 1984, China's Ministry of
Foreign Trade and Economic Cooperation (MOFTEC) authorized the

creation of additional FTCs at the provincial and local levels to encourage greater competition in foreign trade. By 1988, the number of FTCs had increased to about 5,000 at all levels of government; however, that number was reduced to ~4,000 in the 1990s after eliminating and consolidating the FTCs that did not conform to national foreign trade policy. In addition to FTCs, in recent years the government has granted the right to conduct foreign trade to a large number of large and medium state-owned enterprises. Therefore, the number of Chinese trading enterprises as of 1997 reached over 8,000.

It is increasingly difficult for government administrators to effectively regulate such large-scale economic activities. This is reflected in the significant decline in the government-planned share of total foreign trade. Government-planned imports constituted only 18.5 percent of all imports in 1992, compared with more than 90 percent of total imports at the beginning of the 1980s; the planned share of exports was only 45 percent in 1988, and almost all government export planning has been abolished since 1991. Nevertheless, the state still retains some control through licensing, which must be approved by and registered with MOFTEC. A tax of 10 to 20 percent (depending on the technology involved and the existence of an applicable bilateral tax treaty) is withheld on royalty payments.

In an effort to decentralize and liberalize governmental decision-making regarding international trade, the State Commission for Economic Restructuring announced in 1997 that more production enterprises will have the independent power to conduct import and export business, as will some enterprises of domestic trade. China has also abandoned subsidy systems in foreign trade, except for various indirect subsidies on energy, raw materials, and others. Restrictions on some large-scale exports and imports of commodities are to be eased gradually. In addition, other foreign trade regulations such as import substitution, administrative examination and approval, and restrictions on trading rights of foreign investors are being gradually phased down.

2.2.3 Major Trade Regulations

The Ministry of Foreign Trade and Economic Cooperation and other relevant departments under the State Council may draft and issue the catalogues of goods and technologies whose import or export is restricted or banned. Prohibitions or restrictions on imported and exported goods and technologies can only be made according to China's Foreign Trade Law, which is consistent with GATT and WTO agreements. Goods whose import or export is restricted are subject to quota or license management; technologies are subject only to license management. Goods and technologies subject to quota or license management can be imported or exported only when approved by the relevant departments.

Through administrative decrees and regulations MOFTEC regulates and controls license systems for foreign trade enterprises, the trade agency system, dumping procedures, import and export chambers of commerce, border trade, and so forth. China's 1996 Foreign Trade Law governs issues of both commodity and technology trade but can only lay down a few broad principles regarding international trade in services. As the fundamental law in the area of foreign trade, the Foreign Trade Law prescribes the procedures for foreign trade enterprises to import and export most goods and technologies. It also dictates that quotas and licenses must be used to control items that have to be prohibited or restricted in trade, and that open and fair competition must be applied in the allotment of quotas. Other trade regulations include tariffs, import taxes, and licenses.

2.2.3.1 Tariffs and Import Taxes

The most extensive guide to Chinese Customs regulation is the Official Customs Guide, compiled by the Customs General Administration (CGA). In addition to assessment and collection of tariffs, the CGA collects a value-added tax (VAT), generally equal to 17 percent, on imported items. Certain imports are also subject to a consumption (excise) tax. Import tariff rates are divided into two categories: the general tariff and the minimum (most-favored-nation) tariff. U.S. imports are assessed the minimum tariff rate, since the United States has an agreement with China containing reciprocal preferential tariff clauses. China's five Special Economic Zones, open cities, and foreign trade zones offer preferential duty reduction or exemption.

2.2.3.2 Import Licenses/Quotas

China administers a complex system of non-tariff trade barriers, which includes (1) individual quotas on imports of machinery, electronic equipment and general goods like grain, fertilizer, textiles and chemicals; (2) automatic and non-automatic import license requirements on a smaller number of goods; and (3) a tendering system applied to both quota and non-quota commodities, including machinery and electronic equipment. The nation also reserves 350 line items for designated foreign trade corporations and has other nontransparent administrative controls or restrictions on importation of goods and agricultural products. The U.S.-China market access memorandum of understanding (MOU) signed in 1992 committed China to curtail most of these barriers by 1997. Import quotas for machinery and electronic items, as well as carbonated beverages, are set by the State Economic and Trade Commission under the State Council, while the State Planning Commission administers quotas for a variety of general commodities.

The Commodity Inspection Law stipulates that all goods included on a published inspection list (or subject to inspection pursuant to other laws and regulations or the terms of the foreign trade contract) must be inspected prior to importation, sale, or use in China. The State Technical Supervision Bureau, which is responsible for standard-setting for domestic production, also imposes safety certification controls over certain electrical products such as leakage protectors, insulating wire and cable, power-driven tools, refrigerators, electric fans, air conditioners, televisions, radio receivers, and tape recorders. International Electrical Committee standards have also been adopted for these products.

2.2.4 Policy Instruments to Promote International Trade
China's policies for encouraging foreign trade include the following:

- *Regional targeting of specific areas for export promotion and foreign investment.* The government has designated five special economic zones (SEZs), 14 coastal cities empowered to exercise the same policies as the SEZs, and a series of zones along the Chinese coast connecting the SEZs and open cities to form a coastal opening belt; these offer attractive incentives for foreign investment and trade (see section 2.3)
- Sectoral targeting of certain export industries via three major vehicles:
- A government-sponsored export network in operation since 1985, that includes several hundred factories nationwide producing a range of products; participants receive guaranteed supplies of electrical power, raw material, tax reductions on inputs, and attractive purchase prices
- Special investment funds made available by the Ministry of Foreign Trade and Economic Cooperation for the technological upgrading of selected enterprises; an estimated 1,300 enterprises in the machinery and electronics sector have benefited from such funds [Ma 1997a].
- A favorable exchange retention (FER) policy, now defunct, considerably increased the percentage of foreign exchange earnings a favored exporting enterprise was allowed to retain; when China moved from the fixed exchange rate regime to a managed floating regime in January 1994, the FER policy was abolished, but for a decade the policy particularly favored the machinery and electronics sectors
- *Incentives for foreign-invested exporting enterprises.* Twenty-two regulations published by the State Council in 1986 provide additional incentives to foreign-invested enterprises engaged in exports. Preferential treatment is given to Sino-foreign joint ventures if they are categorized as either export-oriented or technologically advanced projects; for example, enterprises that export 70 percent or more in value of their products may reduce their income tax liability by half at the end of the tax reduction or exemption period.
- *Exemption of customs duties for exporting enterprises.* In 1984, the Chinese government adopted policies to promote manufactured exports based on assembly-type operations. Local enterprises may be exempted from import duties on raw materials provided by overseas suppliers to meet export contracts or for use in manufacturing exports.
- *Export credits.* The People's Bank of China (PBOC) offers several types of trade credits and loans to domestic enterprises and foreign buyers of Chinese exports. First, PBOC offers trade credits in domestic currency to exporting enterprises (most of them FTCs) to finance exports. Second,

PBOC offers export seller's credits to Chinese enterprises selling electronic and machinery equipment in the international market. Finally, in 1992 the State Council approved the creation of a new system of credits (in foreign currency) to be extended to buyers of complete sets of Chinese-made machinery and electronic equipment valued at a minimum of one million dollars per transaction.

- *Exchange rate devaluation.* Before China formally adopted the managed floating exchange rate system, the Chinese foreign exchange system was two-tiered, with an official rate typically held constant for extended periods and a secondary market rate determined by supply and demand. The official exchange rate was substantially overvalued relative to the secondary market rate. Between 1989 and 1993, the difference between these two rates has been reduced via several devaluations of the official exchange rate and a related appreciation of the secondary market rate. This policy has boosted China's exports by reducing the implied export tax rate.

2.3 SPECIAL ECONOMIC ZONES

Five special economic zones, 14 "open coastal cities," the Shanghai Pudong New Area, and numerous other special zones and industrial parks have been established in China to promote export, restructure and upgrade industries, and attract foreign investment. These economic zones enjoy relatively dynamic economic and technological operations, advantageous locations, and preferential policies. The various zones are experimental and still evolving, and there is a good deal of overlap between them, leading to some confusion in defining the different types. However, the basic policy behind the special zones remains consistent: to promote acquisition of and domestic capability in advanced technologies that can be used in both industry and agriculture in practical applications, and to entice foreign investment and expertise to help speed up China's modernization process.

2.3.1 Special Economic Zones
In July 1979 the Central Committee of the Communist Party of China (CCP) and the State Council agreed to experiment with Deng Xiaoping's recommendation to establish special economic zones. Four special economic zones (SEZs) were established that year, and a fifth in 1988:

- the Shenzhen, Shantou, and Zhuiai SEZs, located in Guangdong Province adjacent to Hong Kong and Macao
- the Xiamen SEZ, located in Fujian Province opposite Taiwan
- the Hainan Province SEZ, approved in 1988 as the largest special economic zone in China

China's SEZs provide access to cheaper labor and land and offer better incentive packages to foreign investors than other specialized economic areas in Asia, the major incentive being lower taxes. SEZs offer a 15 percent rate of income tax, compared to the standard 33 percent rate for foreign investment in non-open areas. Export-oriented high-tech projects and investments of more than $5 million enjoy even lower tax rates. Income tax on foreign-invested enterprises is exempted for the first two profit-making years. The SEZs also offer exemptions from import duties for production inputs and income tax on profits if reinvested. In addition to providing the same preferential policies enjoyed in the other four SEZs and 14 coastal cities, Hainan provides additional tax concessions for investment in certain infrastructure projects. Foreign investors can also purchase stocks and bonds and lease state-owned enterprises [Ma 1997a].

The SEZs have become major magnets for international investment and trade because of their proximity to overseas markets and their preferential government policies. The five SEZs together generated 215.5 billion yuan ($26 billion) or 3.2 percent of GDP in 1996. This figure represents an annual average increase of 36.8 percent since 1991. Also in 1996, foreign direct investment covered by contracts and agreements amounted to $6.3 billion, and the combined volume of the five SEZs' imports and exports reached $59 billion, one-fifth of the national total. Overall, by the end of 1996 the five SEZs had attracted an aggregate $26 billion in direct foreign investment, one seventh of the total China had received since its open door policy was implemented in 1979 [IOSC 1997a].

2.3.2 Open Coastal Cities

In 1984, 14 coastal cities were empowered by the State Council to practice almost the same policies as the SEZs to attract overseas investment. These cities are Dalian, Qinhuangdao, Tianjin, Yantai, Qingdao, Lianyungang, Nantong, Shanghai, Ningbo, Wenzhou, Fuzhou, Guangzhou, Zhanjiang, and Beihai. All can directly approve investment projects of less than $5 million. Additionally, Dalian is authorized to approve projects costing up to $10 million, and Tianjin and Shanghai are authorized to approve projects costing up to $30 million [Ma 1997a]. The coastal cities also have the authority to approve matters pertinent to equipment imports and send delegations abroad regarding foreign investment projects. The tax incentives used in SEZs have been extended to the Economic and Technological Development Zones and other areas of the 14 coastal cities (see below), but other areas of the cities have only limited tax incentives.

The overall economic performance of these cities has been satisfactory, although individual performance differs significantly. The annual growth in industrial output from 1984 and 1990 varied from 6.7 percent in Shanghai to 36 percent in Beihai. The export/GNP ratio in the 14 cities was five times higher than the national average of 0.17 in 1990. In addition, the 14 coastal cities utilized 36 percent of China's total foreign investment and exported 58 percent of the country's total exports in the same year [Ma 1997a].

2.3.3 Economic and Technological Development Zones

In 1984-5, the Chinese government redesignated the 14 open coastal cities as "economic and technological development zones" (ETDZs) [DOC 1998]. After over ten years' development and construction, about 200 such zones have opened in municipalities across the country, including in inland provinces, under both State and local oversight. ETDZs mainly attract emerging high-tech industries, which account for about two-thirds of the businesses operating in ETDZs. They provide excellent environments for scientific research and technological development and production. Some ETDZs have established scientific and technical "incubators," "scientific and technological pioneering centers," and risk funds for scientific research.

In the State ETDZs, production-oriented enterprises with foreign investors are subject to Enterprise Income Tax (EIT) at a reduced tax rate of 15 percent. Those scheduled to operate for a period of 10+ years are exempt from the EIT for the first two profit-making years, and from half of the tax in the ensuing three years. After this period, confirmed export-oriented enterprises whose output of export products amounts to 70% or more of the value of their products are entitled to pay an EIT of only 10%. Confirmed enterprises of advanced technology pay only half the EIT for an additional three years. Municipalities may offer additional inducements [OSEZ 1997].

The gross industrial product of the first coastal ETDZs rose from 300 million yuan in 1986 to 135 billion yuan in 1996. In addition, during 1996 more than 100 projects valued over $10 million were initiated in four major ETDZs: Guangzhou, Tianjin, Dalian and Minxing (OSEZ 1997).

2.3.4 High Technology Development Zones/Torch Zones/Other Zones

With less success than hoped for in diffusing the advanced technologies developed and practiced within SEZs and ETDZs, Beijing initiated new "high technology development zones" or "Torch" zones in 1988. These specifically focus on linking technology with production and especially on commercializing indigenous new technologies and research [DOC 1998]. The State Science and Technology Commission is the implementing authority. Similar zones have been established by provincial and municipal governments as well. Key technological fields developed under the program include new materials, biotechnology, electronics and mechatronics, and information, energy saving, and environmental protection technology. Torch zones are similar to the technology parks of other countries. The emphasis is on exporting products. By 1998 there were 53 Torch zones, all located in areas surrounding research institutes and manufacturing entities or within existing special zones such as open ports, SEZs, or ETDZs. About 10,000 Torch projects have been completed. Spurred by various preferential policies, many U.S, Japanese, European and other foreign companies had invested $802 million by 1995, with registered capital of $670 million.

China's ongoing experimentation with special commercial zones includes establishment of new free trade zones in Pudong, Tianjin, and

Shenzhen that have the least restrictive rules for foreign trade and investment as long as products are destined for export [DOC 1998], and bonded zones, small customs supervision districts that unite export-oriented process and foreign trade and have special tariffs. Most of these have been established in coastal areas. By the end of 1996, construction of infrastructure facilities in all bonded areas was valued at $12.6 billion. Nearly 15,000 enterprises had been approved, with combined registered capital of $24.1 billion; of these, 7,538 were foreign-funded, which together involved $8.26 billion in contracted investment [IOSC 1997a].

One special zone receiving much attention is the Pudong New Area of Shanghai, established by the Chinese government in April 1990 adjacent to Shanghai's urban districts on the east side of the Huangpu River. Covering 522 square kilometers and having a population of more than 1.4 million, it is by far the largest of China's special investment zones, and the home of all four of Shanghai's national development zones (Lujiazui Finance & Trade Zone, Zhangjiang High-Tech Park, Waigaoqiao Free Trade Zone, and Jinqiao Export Processing Zone). Pudong is expected to become Shanghai's financial and high-technology district, lead the economic development of the Yangtze River Valley, and rejuvenate Shanghai's status as China's finance and trade center.

Overseas companies operating in Pudong New Area enjoy the same tax, tariff, and licensing privileges as those in SEZs; infrastructure investments (airports, roads, railways, and electricity) enjoy further tax exemptions and reductions; and foreign investment in the services sector (financial organizations, department stores, supermarkets) is allowed. The Shanghai area is also permitted to operate securities markets and issue stocks.

The economic performance of Pudong New Area has been remarkable. It contributed 17.7 percent of Shanghai's GDP in 1996, compared with 8.1 percent in 1990. By 1997, 4,303 foreign-funded industrial, financial, commercial, trade, and real estate development projects had been started there, involving $18.9 billion in contracted capital investment [IOSC 1997].

2.4 FOREIGN DIRECT INVESTMENT

Since the beginning of the 1980s, numerous techniques have been employed to improve the investment environment for foreign enterprises seeking to establish factories in China. Following Deng Xiaoping's trip to South China in 1992, China has undertaken a series of experiments to allow limited foreign participation in service sectors and has placed even greater importance on attracting high technology and infrastructure investment.

2.4.1 Foreign Investment Status

China absorbs foreign capital in two forms: foreign loans and foreign direct investment (FDI). As of June 1996, there were more than 270,000 foreign-invested enterprises operating in China. In that year a total of 24,673 foreign investment projects were approved in China, with an agreed capitalization of $81.6 billion, and a dollar amount of $42 billion. Total foreign capital utilized amounted to $55.3 billion. FDI in 1997 was estimated to be $45 billion.

Table 2.3 shows the steady growth in utilized FDI in this decade. Despite the rise in overall FDI, in 1996 the proportion of total fixed investment fell to $12.7 billion from $13.9 billion in 1995 and $16.7 billion in 1994. Table 2.4 depicts the major sources of FDI in China since 1979. From 1979 to 1996 China actually utilized $283.9 billion in foreign capital, including $174.9 billion in FDI [Qin 1997]. Overall, the focus of foreign investment in China has shifted towards infrastructure facilities, basic and "pillar" industries, and capital- or technology-intensive projects.

Table 2.3
Utilized Foreign Direct Investment in China
(1979-1997, $millions)

Year	Foreign Direct Investment (All Countries*)	U.S. Direct Investment
1979-1989	18,468	1,729
1990	3,410	456
1991	4,366	323
1992	11,008	511
1993	27,515	2,063
1994	33,767	2,491
1995	37,521	3,083
1996	41,726	3,440
1997**	45,177	3,034

*Includes Hong Kong and Taiwan Province; ** Estimated
Source: China State Statistical Bureau (SSB), 1998.

Table 2.4
Contracted Foreign Direct Investment in China,
by Region ($ millions)

Year	Hong Kong	Japan	USA	Taiwan	Others
1979-1989	20,879	2,855	4,057	1,100	4,569
1990	3,833	457	358	1,000	1,948
1991	7,215	812	548	3,430	3,405
1992	40,044	2,173	3,121	5,543	7,241
1993	73,939	2,960	6,813	9,965	17,759
1994	46,971	4,440	6,010	5,395	19,864
1995	40,996	7,592	7,471	5,849	29,374
1996	28,002	5,131	6,916	5,141	28,086

Source: China State Statistical Bureau (SSB), 1997.

2.4.2 General Investment Climate

In general, China's trade, investment, and regulatory system are characterized by a lack of transparency and inconsistent enforcement. A complex and often conflicting system of national, regional, and local administrative controls regulates access by foreign investors to China's market. Foreign nationals and even many Chinese officials lack a solid understanding of China's regulations and of the scope of authority and the duties of various government agencies. Foreign investors in China face such obstacles as limited availability of foreign exchange, the highly personalized nature of conducting business, inadequate protection of intellectual property, absence of a strong contractual and legal tradition, barriers to market access, production controls, unequal treatment compared with domestic companies, and lack of adequate mechanisms for resolving disputes. China has gradually improved its performance on these factors and intends to further improve the investment climate through a series of sweeping reforms, although implementation and enforcement is often difficult.

As noted in section 2.4 above, China's system of investment incentives is elaborate. The investment climate in Special Economic Zones or similar areas is superior to that in other regions, offering lower taxes, better infrastructure, and less bureaucracy, although zones can vary considerably. Incentives are not automatically granted for foreign investors, who must apply and sometimes negotiate for these benefits with relevant governmental authorities on a case-by-case basis. Preference is given to projects involving high-tech and export-oriented investments in priority sectors like energy, transportation, and "pillar" industries such as electronics and machinery.

Potential investment projects usually must go through a multitiered screening process starting with approval of the project proposal. The central government has delegated varying levels of approval authority to local governments and SEZs. Projects exceeding the investment limits

($30 million in special economic zones and $10 million in inland regions) are approved by the Ministry of Foreign Trade and Economic Cooperation (MOFTEC) and the State Planning Commission for new projects, or the State Economic and Trade Commission for projects of existing enterprises. If an investment involves $100 million or more, it must obtain State Council approval in addition to MOFTEC's review and approval. China's authorities prefer investment proposals that promote exports to earn foreign exchange, introduce advanced technology, and provide technical or managerial training. Other proposal evaluation criteria include the fairness of the contract, the percentage of local content, hiring of local employees, whether the technology is available elsewhere in China, and whether China already has sufficient production capacity.

Lack of well developed protection for intellectual property is a disincentive to investment in China. Foreign companies have complained of insufficient patent and software protection, which has inhibited the transfer of advanced technologies. In a January 1992 U.S.-China Memorandum of Understanding on the Protection of Intellectual Property Rights (IPR), China pledged to join relevant international conventions and enact or amend IPR legislation. Nonetheless, enforcement of IPR rights either through judicial or administrative measures remains a serious problem.

Outright expropriation of foreign investment has not occurred in the reform period and is forbidden under current foreign investment legislation, except under special circumstances such as national security considerations and removing obstacles to large civil engineering projects. China is a member of the International Center for the Settlement of Investment Disputes (ICSID) and has ratified the New York Convention on the Enforcement of Foreign Arbitral Awards. However, Chinese authorities place strong emphasis on resolving disputes through informal conciliation. When a formal mechanism is required to settle the disputes, the authorities prefer arbitration through Chinese agencies. Litigation is considered only reluctantly as a final option. Many foreign investors have found the Chinese approach time-consuming and unreliable.

Skilled workers are often in short supply, especially managers and those with marketing skills. In general, foreign invested enterprises (FIEs) are free to pay whatever wage rates they want above a locally designated minimum wage. In addition to basic wages, provision of subsidized services such as housing and medical care is common and constitutes a very large portion of a venture's labor expenses. The rates for payment of overtime compensation under various circumstances are specified in China's National Labor Law. Terminating individual workers for cause is legally possible throughout China but may require prior notification and consultation with the local union. The National Labor Law provides for establishment of collective labor contracts to specify wage levels, working hours, working conditions, and insurance and welfare. China's foreign investment legislation and labor law require FIEs to allow union recruitment, but do not require an FIE actually to set up the union; on the other hand, most coastal provinces have passed stricter regulations that require unions in all FIEs. Still, the majority of such enterprises do not have unions and some (contrary to Chinese law) have agreements with localities not to establish unions in their factories.

2.4.3 Forms of Foreign Direct Investment

Foreign investment is allowed in the Chinese economy primarily in equity joint ventures and cooperative (contractual) joint ventures; there are also some wholly foreign-owned enterprises. The fundamental legislation dealing with foreign investment in China are the Law on Chinese-Foreign Equity Joint Ventures and the Law on Wholly Foreign-Owned Enterprises. Joint, cooperative, and wholly foreign-owned ventures in China must be approved by and registered with the Chinese Government. They are legal entities and subject to both the jurisdiction and the protection of Chinese law in respect to legitimate rights and interests and earnings. They may not engage in any activities detrimental to China's public interests or sovereignty, cause environmental pollution, or impair the rights or interests of any partner, and they must help to develop China's economy and raise the nation's scientific and technological levels. Table 2.5 displays the three major forms of direct foreign investment or ownership allowed in China since 1979 in terms of number of contracts and money invested.

2.4.3.1 Equity Joint Ventures

An equity joint venture takes the form of a limited liability company. Each joint venture partner's liability is limited to the capital subscribed by it. The Joint Venture Law specifies that the proportion of the foreign joint venture's investment in an equity joint venture must be at least 25 percent of its registered capital. According to government policy, joint ventures must comply with at least one of the following requirements:

- use advanced technical equipment and scientific managerial methods that help increase the variety, improve the quality, and raise the output of products, and save energy and materials
- be open to technical innovation so as to bring about quicker returns and bigger profits with less investment
- help expand exports and thereby increase foreign currency receipts
- help train technical and managerial personnel

Joint ventures are allowed in most industries, except those restricted due to strategic or economic reasons. An equity joint venture in China is subject to examination and approval by MOFTEC and must be registered with the State Administration for Industry and Commerce (SAIC) or any of its local and regional bureaus. The whole acceptance process, beginning with identifying a potential partner and ending with the state approvals, takes one to two years or even longer.

Table 2.5
Foreign Direct Investment in China by Forms of Ownership

Year	Equity Joint Venture		Contractual Joint Venture		Wholly Foreign- Owned Enterprises	
	No. of Contracts	Amount ($M)	No. of Contracts	Amount ($M)	No. of Contracts	Amount ($M)
1979-89	12,198	12,530	7,994	13,558	1,525	3,144
1990	4,091	2,704	1,317	1,254	1,860	2,444
1991	8,395	6,080	1,778	2,138	2,795	3,670
1992	34,354	29,128	5,711	13,255	8,692	15,696
1993	54,003	55,174	10,445	25,500	18,975	30,457
1994	27,890	40,194	6,634	20,301	13,007	21,949
1995	20,455	39,741	4,787	17,825	11,761	33,658
1996	12,628	31,876	2,849	14,297	9,062	26,810
1997*	8,786	19,343	2,251	11,936	9,282	16,068

* Estimated. Source: China State Statistical Bureau (SSB) 1998

2.4.3.2 Cooperative Joint Ventures

Cooperative (contractual) joint ventures are more flexible and easier to establish than equity joint ventures. Chinese and foreign investors can mutually agree on such matters as conditions for cooperation, distribution of earnings, sharing of risks and costs, methods of operation and management, and ownership of assets at termination of the agreement. The cooperative parties are jointly liable for all debts of a cooperative joint venture; therefore, they must present guarantees from a bank or a parent company that they will fulfill their joint obligations. Most contractual joint ventures are registered as Chinese legal entities and thus only assume liability for debts within their properties. The proportion of investments contributed by the foreign partner must be at least 25 percent of the registered capital.

In general, China encourages cooperative joint ventures in industries such as energy, transportation, raw materials, electronics, environmental protection, machine manufacturing, agriculture, forestry, animal husbandry, tourism, and service trades. Government restrictions and requirements for approval are similar to those for equity joint ventures. Depending on the total value of a given project, whether the domestic partner is a government entity, and whether the country has to consider the project's impact on fuel, power, communications, and transportation, the joint venture can be examined and approved by the Economic and Trade Administration Department of specific regions where the joint venture is to be established (e.g., province, autonomous region, municipality, or Special Economic Zone); by MOFTEC; or by the relevant ministry, committee, or bureau of the State Council. After receipt of an approval certificate, the partners must register with the SAIC, the tax authorities, and the labor bureau where the cooperative joint venture is located.

2.4.3.3 Wholly Foreign-Owned Investment Enterprises

Foreign investors in a wholly foreign-owned investment enterprise establish the enterprise in China with their own capital and in accordance with relevant Chinese laws. Such an enterprise can be registered as a Chinese legal entity in the form of a limited liability company. It performs production and operation activities independently, subject to examination and approval and the issue of a business license. The Law on Wholly Foreign-Owned Enterprises establishes principles and rules governing the legal status of the enterprise; conditions for, scope, and founding procedures of the enterprise; and support, management and operation of the enterprise.

Achieving a balance in foreign exchange is a basic requirement for operation of a wholly foreign-owned enterprise in China. Both the Law on Wholly Foreign-Owned Enterprises and its rules for implementation emphasize that a wholly foreign-owned enterprise shall achieve by itself a balance of revenues and expenditures in foreign exchange. However, the law also has some flexibility. For example, if an enterprise uses advanced technology, machinery, and equipment to produce goods that were previously imported and is permitted to sell the products in China, but consequently experiences an imbalance in foreign exchange, the relevant authorities will help it to correct the imbalance.

As in joint ventures, the level of governmental authority required to examine and approve a wholly foreign-owned enterprise depends on the total amount of investment and the industry concerned. The foreign investor must submit to MOFTEC and local authorities a feasibility study, articles of association of the enterprise, the name-list of the legal representatives or the board candidates, financial documents of the foreign investor, and an inventory of goods and materials for import. After the application is examined and passed and an approval certificate issued, the foreign investor must file an application for registration with the relevant administrative department for industry and commerce, and must obtain a business license.

Wholly foreign-owned investment enterprises have the advantage of being entirely controlled and managed by the foreign owners; nevertheless, many foreign investors prefer to have a domestic partner to help them understand the local market situation and deal with problems that the Chinese partner is better able to handle.

2.4.4 Foreign Investment Policy in the Electronics Industry

The Chinese government announced new "Provisional Regulations on Guiding Foreign Investment" and "Guidelines on the Industrial Catalog for Foreign Investment" in October 1995. The regulations and guidelines, developed jointly by the State Planning Commission, the State Economic and Trade Commission, and the Ministry of Foreign Trade and Economic Cooperation, specify the types of projects for which China encourages, restricts, and prohibits foreign investment. Foreign investment was encouraged for projects involving energy, communication, important raw and processed materials, and new agrotechnology. China encourages foreign investment in the manufacture of the following electronics products:

- large-scale integrated circuits (LSICs)
- new electric and electronic components (including IC chips)
- photo-electronic devices and sensors
- mainframe and mini-computers
- top-end personal computers with 32 bits or higher CPUs
- 900 MHz digital cellular mobile communication equipment
- DS-5 or higher SDH fiber-optic communication equipment and network management equipment.
- digital microwave communications systems and measuring equipment
- asynchronous transfer mode (ATM) switching equipment
- mercury-free alkaline manganese batteries, lithium batteries, and hydrogen-nickel batteries
- key components for facsimile equipment (such as heat-sensing printing heads and picture-sensing elements)
- digital magnetic tape recorders and players for compatible digital television and high definition television (HDTV), and laser disc players
- satellite communications terrestrial earth stations (TES) and data earth stations (DES) and their key components
- Commercial satellites, satellite payloads, and satellite applications

China also encourages foreign investment in the development and production of the following:

- new semiconductor and photoelectronic materials
- software (including computer and telecommunications software)
- new fields — information and communications networking technology, international economic, scientific, and technological information services, and the like

Foreign investment is restricted for production/manufacturing of specific electronics and telecommunications components, products, and equipment:

- radio cassette players and radios
- black-and-white television sets
- personal computers with 16 bits and below CPUs (including 16 bits)
- radio telephone equipment with frequency bands of 450 MHz and below
- radio and television broadcasting systems
- color television sets and tuners and remote controls

- CRT and glass shells for color TV sets
- video cameras (including camcorders and VCRs)
- magnetic heads, magnetic drums, and video recorder core components
- analog mobile communications system (cellular handsets, pagers, and cordless phones)
- fax machines
- satellite television receivers and key components
- microwave relay communications equipment below 140 Mbps rate
- central office switches and private automatic branch exchanges (PABXs)

Foreign investors are prohibited from investing in the following areas:

- operation and management of posts and telecommunication services
- operation of radio and television stations at any level (including cable television networks, broadcasting stations, and relay stations)
- production, publication, and distribution of radio or television programs
- construction of telecommunications projects that jeopardize the safety and effectiveness of military installations

2.5 FINANCIAL SYSTEM AND FOREIGN EXCHANGE MARKET

Before 1983, China's financial system operated through the State collecting revenue from state enterprises and allocating investment through budgetary grants. Fixed asset investments in SOEs were all direct transfers or grants from the government budget. The State provided the credit needed by enterprises to implement plans for physical output and provided and monitored cash used to cover labor costs and purchases of agricultural products. In 1983, direct grants to agriculture, construction, and production enterprises were replaced with interest-bearing loans, and the banking system since then has gradually become the primary channel through which investments are financed and the central authority exercises macroeconomic management. In 1994, the government inaugurated further monetary reform emulating Western-style monetary control systems and stressing the goals of stabilizing the currency and supporting economic growth.

2.5.1 Monetary Policy

Controlling inflation and keeping exchange rates and currency values stable remain the ultimate objectives of Chinese monetary policy. Table 2.6 displays China's major financial indicators since 1991. The three major indicators of money supply, newly printed yuan (M0), cash, demand deposits, and checks (M1), and time deposits plus M1 (M2), grew 16-17 percent in 1997. The official exchange rate between Chinese yuan and U.S. dollars has changed very little since 1994; domestic debt has increased significantly.

Table 2.6
Financial Indicators (US $ Billions)

Year	M0 Supply (1)	M1 Supply (2)	M2 Supply (3)	Official Exchange Rate (Yuan/$)	Domestic Debt (Treasury Bond Issues)	Foreign Debt
1991	59.7	162.3	363.7	5.32	3.7	3.4
1992	78.6	212.5	460.2	5.52	7.2	3.8
1993	101.8	282.6	605.6	5.76	5.5	6.2
1994	84.6	238.3	544.4	8.62	11.9	1.7
1995	99.5	287.1	697.4	8.35	18.1	0.5
1996	111.7	343.5	916.8	8.30	22.3	1.4
1997*	129.6	401.6	1,079.3	8.27	29.2	N/A

* Year-end estimation; (1) newly printed yuan; (2) cash, demand deposits, and checks; (3) time
deposits plus M1.
Source: China State Statistical Bureau 1998.

2.5.2 Monetary Management

The People's Bank of China (PBOC), which is a ministry-level agency
under the direction of the State Council, serves as China's central bank. It
helps to formulate and then implements China's monetary policy and ensures
the compliance of all financial institutions with State policy. Its duties
include (1) drawing up and implementing monetary and interest rate policies;
(2) directing and supervising banks, non-bank financial institutions, and
insurance companies; and (3) examining and approving the establishment,
merger, and dissolution of financial institutions, insurance companies, and so
forth. There are six regional PBOC branches.

The PBOC continues to follow a tight monetary policy in order to
upgrade cash management and control the total amount of currency in
circulation. Since cash flow constitutes a fairly large share of the basic
currency composition in China (~33 percent in 1996), controlling the amount
of currency in circulation is vital. In response to China's macro-economic
control policy of 1993-96, the inflation rate fell from 21.7 percent (retail
price index) to 0.8 percent in 1997 while the economy grew at an annual rate
of close to 10 percent (see Table 2.1).

Properly managing loan increments and strictly controlling the credit of
State banks are two other important tasks of the PBOC. It increases the
money supply as needed by providing loans to various industrial sectors,
including working capital loans to viable enterprises with marketable
products; credit support to projects under construction that promise good
returns; and loans for science and technology development. It also uses credit
policy to help adjust the structures of money supply and demand: with regard
to demand, the PBOC grants loans for housing projects and home mortgages;
with regard to supply, it organizes syndicate loans, cultivates enterprise
groups, and supports strategic restructuring of large enterprises.

The PBOC utilizes a number of policy instruments to manage the monetary system:

- *Credit planning.* The PBOC determines the credit ceiling, i.e., total loans, that can be extended within a year for each specialized bank and each of its local branches. The overall credit plan is determined by the PBOC (which formulates the total amount of the money supply) and its provincial branches (which formulate provincial plans for deposits, loans, and cash issuance). Credit planning has been the major monetary policy instrument, although its importance has gradually diminished.

- *Setting reserve ratios.* The specialized banks must submit a certain proportion of their deposits to the central bank in the form of required reserves. The effective reserve ratio was 20 percent in 1992 (the official ratio of 13 percent plus an additional excess reserve ratio of 7 percent).

- *Adjusting interest rates.* The PBOC controls interest rates for deposits and lending in all specialized banks. In general, government lending to agricultural, infrastructure, and energy sectors is subject to lower rates, in accordance with industrial policy. The effectiveness of this policy instrument is limited since official rates do not reflect market conditions, and state-owned enterprises are often insensitive to interest rates.

- *Lending to commercial banks.* The PBOC extends credit to banks that fall short of funds for meeting the reserve requirement or for local lending. The lending rate for such credit is often used as an instrument for controlling the money supply.

- *Ensuring compliance with the Law of the People's Bank of China* (the PBOC Law). This law passed by the People's National Congress in 1996 authorizes that policy instruments to conduct monetary policy may include rediscounting, open market operations, trading of foreign exchange, and other instruments prescribed by the State Council.

The PBOC is to gradually phase down use of the mandatory credit plan and increase use of more indirect means to control the money supply, such as open market operations, central bank lending, discount rates, reserve ratios, interest rate adjustments, and operations in the foreign exchange market. Financing through sale of government bonds and through printing money are strictly prohibited. The maximum length of maturity for interbank loans has been shortened from three months to three days.

To facilitate open market operations, the PBOC issued a number of short-term central bank bonds in 1995 and 1996. In January 1996, the PBOC lifted controls over interest rates in the interbank market, and it has adjusted the official interest rates frequently to reflect market conditions. Loans from central banks can be recalled if the PBOC considers banks' liquidity too high, and vice versa. This policy instrument has gradually played an important role in liquidity management.

2.5.3 Banking System

Since the early 1990s, China's banking system has been restructured to create a more modern and international system in China and one that is able to more efficiently allocate financial resources. The former State banks are now commercial retail banks; non-bank financial institutions have begun to operate and compete with State banks for savings and loans; various specialized banks have been created or reestablished, becoming the main channels of enterprise finance during the 1980s; and many foreign banks, insurance companies, and other financial institutions have been allowed to operate in China with a limited scope. For example, on December 31, 1996, the PBOC officially licensed Hong Kong Bank, Citibank, Bank of Tokyo, Mitsubishi, and the Industrial Bank of Japan to conduct business in yuan in the Shanghai Pudong New Area on a trial basis. On January 24, 1997, the Standard Chartered Bank, Sanwa Bank, Daiichi Kangyo Bank, and the Shanghai-Paris International Bank were also granted licenses to offer yuan deposit and loan services to China-based foreign businesses in the Pudong area. It is expected that more foreign financial institutions will be allowed to conduct yuan business in the future.

China's banking system comprises a wide variety of "specialized" banks under the PBOC, which were once integral elements of the socialist economic planning system, functioning as conduits for State investment: the Agriculture Bank of China, the Bank of China (not to be confused with the PBOC), the People's Construction Bank of China, the Industrial and Commercial Bank of China, and the Bank of Communications. Under the 1994 monetary reform plan, these are all evolving into independent commercial banks operating on market principles.

In a separate category of banks under the supervision of the PBOC are three new "policy" banks established in 1994 to grant policy loans to selected projects: the State Development Bank of China, the State Agricultural Development Bank, and the Import-Export Bank of China. These policy banks have the financial responsibility for financing economic and trade development and state-invested projects included in the government's five-year economic plans. Their sources of capital are mainly government budgetary, social insurance, postal, and investment funds.

Overall, China's monetary and banking reform is far from complete. A serious problem is that some of the commercial banks are still burdened by a stock of bad debt resulting from past policy lending and thus could face a solvency crisis in the future. The PBOC's ability to use indirect policy instruments to conduct macroeconomic management such as open market operations is still very limited; in addition, commercial banks are still overburdened by the large volume of non-performing loans made to state-owned enterprises.

2.5.4 Foreign Exchange Reform

Before 1978, the exchange rate played no direct role as a price signal in either foreign trade or in the allocation of resources in China. All foreign trade was carried on according to mandatory import and export plans determined by the central planning authority in an attempt to achieve balance in foreign exchange requirements. The reliance on mandatory plans implied that the foreign exchange rate had very little influence on the level and the pattern of foreign trade.

After 1978, FTCs were given a certain degree of freedom to conduct foreign trade. The exchange rate began to act as a market signal and impacted decisions on foreign trade. However, because the domestic currency was overvalued and the government required the FTCs to surrender all their foreign exchange earnings at that overvalued exchange rate, they had little incentive to expand exports. Shortly after decentralizing foreign trade, the government of China began to modify its foreign exchange policy in order to promote exports and establish a market for swapping foreign exchange, primarily using three methods:

- *Devaluing the domestic currency to reflect market value:* official exchange rates have been steadily devalued, beginning with a 44.7 percent devaluation on January 1, 1981, and continuing with six devaluations from 1981 to 1993 ranging from 9.6 to 44.9 percent.

- *Introducing the foreign exchange retention program:* starting in 1978 exporting enterprises and local governments were allowed to retain a percentage of their foreign exchange earnings to finance their own imports; the percentage increased from about 44 percent from 1987 to 1990 to 80 percent in 1993 (World Bank 1994; Wong and Wong 1997). This was a transitional policy and has since been discontinued.

- *Permitting exporting enterprises and local governments to sell foreign exchange quotas* to units that seek access to foreign exchange in order to purchase imports. The Guangdong Branch of the Bank of China established the first foreign exchange swapping service in 1980. Later on, foreign exchange swap markets opened in Shenzhen (1985), in Shanghai and Beijing (1986), and in Tianjing (1987).

Although the retention scheme and the swap markets were not entirely free market mechanisms and were subject to various restrictions, they did help to increase the convertibility of China's domestic currency.

On January 1, 1994, China unified the two-tier exchange rates at the prevailing swap rate of 8.7 yuan per U.S. dollar. A national interbank foreign exchange market replaced foreign exchange swap markets, and a foreign exchange selling and buying system was introduced. This reform was a major step towards achieving current account convertibility. Under this new "managed floating" system, exchange rate stability and the current account balance have to be maintained through the PBOC's monetary policies and

MOFTEC's administrative controls rather than through foreign exchange controls on the current account transactions. Consequently, exchange rate determination has had a much broader effect on China's macroeconomic performance than simply managing imports and exports.

On December 1, 1996, the PBOC introduced full convertibility of China's currency for current account (trade) transactions. The move marked an important step forward on currency convertibility, though China still restricts convertibility on its capital account. China lacks a foreign exchange market where foreign exchange dealers can interact directly with international markets. The restrictions can be justified by the fact that foreign exchange reform took place prior to reforms in other sectors of the economy; in addition, since state-owned enterprises were still not subject to hard budget constraints and the banking sector had not been completely commercialized, immediate removal of all controls could have caused massive devaluation and created severe macroeconomic instability. However, restrictions have to be gradually phased out so as to improve the efficiency of the interbank foreign exchange market.

2.6 FISCAL STRUCTURE

Over the last twenty years, China has replaced its traditional revenue remittance system with a Western-style tax system and reduced the scope of government involvement in the production sector. In addition, the government has decentralized the fiscal management system by granting localities more flexibility in collecting revenues and deciding expenditures.

2.6.1 Fiscal Structure

Before economic reform began, China's government revenue and expenditure each accounted for nearly one-third of the national GDP. Both government revenue and expenditure as a percentage of GDP have steadily declined and were about 11 percent in 1995 (Table 2.7).

Several factors have contributed to the decline in the government sector's share of GDP: (1) authorization during the first few years of reform for enterprises to retain some of their profits; (2) state sector failures to contribute to government budgetary increases; (3) disincentives to local tax efforts due to the nature of central-local fiscal relations [Wong, Heady, and Woo 1995; Ma 1997b].

Table 2.8 displays sources of government revenue for 1994. Revenues collected through value-added tax accounted for almost 42 percent of total revenue. Of total revenue, remittances from enterprises constituted about 1.4 percent ($900 million or 7.8 billion yuan) compared to more than 40 percent (43.5 billion yuan) in 1980.

Table 2.7
Government Revenue and Expenditure in China, 1990-1995 ($Billions)

Year	Revenue	Expenditure	Deficit	Revenue/GDP Ratio (%)	Deficit/GDP Ratio (%)
1990	61.4	72.2	10.8	15.8	2.8
1991	59.2	71.7	12.5	14.6	3.1
1992	63.3	79.5	16.3	13.1	3.4
1993	75.5	91.8	16.3	12.6	2.7
1994	60.5	73.1	12.6	11.2	2.3
1995	74.8	81.7	7.0	10.7	1.0

Source: China State Statistical Yearbook, 1996.

Note: Borrowing is not included in revenue. Debt repayment is included in expenditure. Exchange rates based on China's official figures: 1990 (4.78), 1991 (5.32), 1992 (5.52), 1993 (5.76), 1994 (8.62), 1995 (8.35), and 1996 (8.30).

Table 2.8
Source of Government Revenue in 1994

Revenue Category		$Billions*	%
Tax	Value Added Tax	26.8	41.6
	Business Tax	7.8	12.1
	Consumption Tax	5.6	8.8
	Income Tax from State-Owned Enterprises (SOEs)	7.1	11.0
	Income Tax from Collective-Owned Enterprises (COEs)	1.1	1.8
	Customs Duties	3.2	4.9
	Agricultural and Animal Husbandry Tax	2.7	4.2
Revenue Remittances from Enterprises		0.9	1.4
Funds for Energy and Transportation Key Projects		0.6	1.0
Budget Adjustment Fund		0.7	1.1
Other Revenues		7.0	10.8
Subsidies to Loss-making Enterprises		-4.2	-6.6
Total Revenue		64.4	100.0

* Official exchange rate $1=8.62 yuan

Source: Finance Yearbook of China 1995

Table 2.9 (Continued on next page)
Government Expenditure, 1978 and 1995 ($Millions)

Expenditure Category	1978	%	1995	%
Capital Construction	16,142.9	40.3	9,449.1	11.6
Circulating Funds	2,392.9	5.9	419.2	0.5
Technical Upgrading	2,250.0	5.6	5,928.1	7.3
Geological Prospecting	714.3	1.8	790.4	1.0

Expenditure Category	1978	%	1995	%
Administration of Industry and Commerce	642.9	1.6	1,233.5	1.5
Agriculture Development	2,750.0	6.9	5,149.7	6.3
Cultural, Education, Science and Health	4,035.7	10.0	17,568.9	21.5
Pension and Social Welfare	642.9	1.6	1,377.2	1.7
National Defense	6,000.0	15.0	7,628.7	9.3
Government Administration	1,750.0	4.4	10,455.1	12.8
Price Subsidies	392.9	1.0	4,371.3	5.4
Repayment of Debts	n/a	n/a	10,622.8	13.0
Total	40,071.4	100	81,724.6	100

Based on China's official exchange rates: 1978 ($1=2.80 yuan) and 1995 ($1=8.35 yuan)

Source: China State Statistical Yearbook, 1996.

Table 2.9 compares the functions of government expenditure for the years 1978 and 1995.   Since 1980, the composition of government expenditures has changed in three major areas:

- investment in capital construction has declined significantly, falling to 11.6 percent in 1995

- expenditure on social welfare, education, culture, and health has increased — from 13.1 percent in 1978 to 23.2 percent in 1995

- expenditures on administrative expenses, price subsidies, and debt repayment has increased significantly

Declining government expenditures on economic construction reflects the declining importance of China's state-owned enterprises in total output. Increases in social expenditures are consistent with government policy to shift away from direct involvement in private economic activities, as well as concern over deteriorating terms of income distribution. Finally, debt financing has become a policy option and was written into the 1994 budget law. The government budget has been transformed from an instrument of central planning into an instrument of indirect macroeconomic management.

2.6.2 Taxes

Before 1978, there were no personal or enterprise income taxes in China. The government raised revenue through profit remittances from State-owned enterprises (SOEs). In 1979, the government introduced the profit retention system, under which SOEs were allowed to retain a portion of their profits in order to provide incentives to increase profits. However, the system was not standardized, and the government had to frequently revise the retention rates. The profit retention system was replaced by a uniform enterprise income tax system in 1983, which came under criticism for creating an unequal distribution of retained profits across enterprises. In 1987-8, the SOE income tax system was again replaced by the "contract responsibility system," in which SOEs remitted a certain amount of their profits to the government based on individually negotiated contracts. The objectives of the contract responsibility system were to increase the autonomy of SOEs and encourage them to maximize profits, as well as to stabilize government revenue.

The contract responsibility system did not lead to a substantial increase in SOE profitability. The ratio of government revenue to GDP continued to decline, primarily because enterprises were not held responsible for their financial losses. Enterprises did not have sufficient incentive to improve their financial performances because when they suffered losses, the government renegotiated the amount of contracted revenue remittance, increased subsidies, or offered special credits. Another problem was that different types of enterprises (state-owned, collectively-owned, and foreign-invested) were taxed at different rates, based on the form of ownership. In addition, there were many conditions under which an enterprise could be exempted from taxation or enjoy reduced tax rates. To correct these problems, the Chinese government launched a major tax reform in 1994. The major changes in the 1994 tax reform package are summarized below.

2.6.2.1 Turnover taxes

China's value-added tax (VAT) covers all manufacturing, wholesale, and retail enterprises, regardless of whether they are domestic, foreign-owned, or joint venture enterprises. Unlike the consumption-based VAT in many Western countries, the Chinese VAT is based on product origins. For most products the VAT rate is 17 percent. A business tax of 3 to 5 percent is applied to services other than retail and wholesale business, and to real estate sales. A new consumption tax applies to a small number of consumer goods. The product tax and industrial and commercial taxes assessed on foreign-invested enterprises have been abolished.

2.6.2.2 Enterprise income taxes

The income tax rates for large- and medium-sized SOEs have been cut from 55 percent to a uniform 33 percent. The same rate is applied to all other types of enterprises regardless of ownership, although there is a preferential rate of 15-24 percent in areas such as SEZs, ETDZs, and Open Coastal Cities. The income adjustment tax and mandatory contributions to various funds formerly levied on SOEs have been abolished.

2.6.2.3 Personal income taxes

A uniform personal income tax is applied to Chinese and foreigners. The monthly deductible allowance on the personal income tax is 800 yuan (~$96), but additional deductions are allowed for foreigners. A progressive rate from 5 to 45 percent is applied to income from wages and salaries, and a progressive rate from 5 to 35 percent is applied to income from business activities of private manufacturers and merchants and to subcontracting and rental income. A 20 percent flat rate is applied to income from publications, remuneration for services, patents and copyrights, interest and dividends, rental and transfer of assets, and other sources.

China has made progress towards broadening the tax base, increasing social expenditures, and simplifying its tax system. However, many problems must still be addressed: tax avoidance is still a serious problem in the enterprise income tax, many loopholes exist in administration of the individual income tax, and the burden of the VAT to different trades is distributed inequitably. During the 9th five-year plan (1996-2000), China plans to reform the tax system by concentrating on unifying tax laws, equalizing tax liabilities, simplifying the tax system, and rationally dividing authority between central and local tax administrations.

2.6.3 Import Tariffs

China has steadily reduced its import tariffs over the past several years. In 1995, the State Council dropped China's import tariffs to 23 percent on 4,963 categories of commodities and announced that imported equipment and raw materials would also be levied tariffs and import link taxes at that tax rate. A further reduction of import tariffs was announced in 1997, that brought down China's simple average tariff to 17 percent. Table 2.10 summarizes China's reductions of import tariffs since 1992. During the November 1996 Asia-Pacific Economic Cooperation (APEC) forum ministers' meeting, China presented its Individual Action Plan, promising to lower its average tariff to 15 percent by 2000. Table 2.11 compares 1998 tariff rates of China and the United States for various major electronics products.

2.7 U.S.-CHINA TRADE RELATIONS

Trade relations between the United States and China are of growing importance in all Asian and international trade. Controversial issues regarding U.S.-China trade relations include U.S. decisions to grant Most Favored Nation (MFN) trade status to China, China's application to the WTO, China's trade barriers to U.S. exports, and U.S. policy on controlling exports to China.

2.7.1 Trading Status

After a thirty-year hiatus, bilateral trade between the United States and China resumed following establishment of diplomatic relations between the two countries in 1979. Since then, U.S.-China trade and economic relations have experienced a period of rapid growth. Bilateral trade went from $2.4 billion in 1979 to $63.5 billion in 1996 (see Table 2.12). China has became one of the United States' top five trading partners. In 1997, the U.S. trade deficit with China was expected to reach $48.7 billion (see Table 2.11), second only to the U.S. trade deficit with Japan.

Table 2.10
The Arithmetic Average Tariff of China (%)

Year	1992	1993	1994	1995	1996	1997
Arithmetic Average Tariff	43.1	39.9	35.9	35.3	23.0	17.0

Source: Asia System Media Corp. [1997]

Table 2.11
China and U.S. Import Tariff Rates for Major Electronic Products (%)

Description of Goods	China		U.S	
	MFN*	General	MFN*	Non-MFN
Color cathode-ray tubes	20	40	9-15	60
Data/graphic display tubes	12	17	3.6	35
Microwave tubes	12	17	3.6	35
Diodes other than photosensitive or light emitting	12	30	Free	35
Transistor, other than photosensitive transistors	12	30	Free	35
Photosensitive and other semiconductor devices	12	30	Free	35
Light emitting diodes	12	30	0.3	20
Mounted piezoelectric crystals	12	30	Free	35
Cards incorporating electronic integrated circuits	6	10	Free	35
Metal oxide semiconductors (MOS technology)	6	24	Free	35
Other monolithic integrated circuits	6	24	Free	35
Video projectors	6	24	Free	35
Hybrid integrated circuits	6	30	Free	35
Electronic microassemblies	6	30	Free	35
ICs and microassembly parts	5	30	Free	35

* Import tariff rates for most-favored-nation
Source: U.S. Department of Commerce, 1998.

Table 2.12
Sino-U.S. Trade Statistics ($ billions)

Year	Chinese Statistics			U.S. Statistics		
	Chinese Export	Chinese Import	Balance	U.S. Export	U.S. Import	Balance
1993	17.0	10.7	6.3	8.8	31.5	-22.8
1994	21.5	14.0	7.5	9.3	38.8	-29.5
1995	24.7	16.1	8.6	11.8	45.6	-33.8
1996	26.7	16.2	10.5	12.0	51.5	-39.5
1997	N/A	N/A	16.1*	10.3**	52.0**	-48.7*

Source: Chinese Customs and U.S. Department of Commerce
* Forecast based on January-October 1997 data
** January-October 1997 data

Table 2.12 shows that big discrepancies exist between data from the United States and China on the trade deficit. One reason for the discrepancy is that much of China's trade flows through Hong Kong. The U.S. counts everything that moves from China through Hong Kong as a Chinese export; until 1993, China counted it as an export to Hong Kong. Thus U.S. government data overstates the size of the trade deficit with China, and Chinese data understates it [Wessel 1998].

2.7.2 Most Favored Nation Trade Status

The United States grants permanent Most Favored Nation (MFN) trade status, with lower tariffs and enhanced market access, to all of its major trading partners except China. Because China is a "non-market" economy country, temporary, annually renewable MFN status has been granted to China subject to the conditions of the Jackson-Vanik amendment to the U.S. Trade Act of 1974. In response to U.S. granting of MFN status to China, China likewise allows U.S. goods to enter China under low-tariff and high-access conditions.

Between 1980 and 1989, the U.S. process of renewing China's MFN status was routine. However, since the Tiananmen Square incident in 1989, many members of the U.S. Congress have come to view the threat of withdrawing China's MFN status as a major lever to encourage the Chinese government to improve its human rights record. Each year since 1989, bills have been introduced but not enacted that would either terminate or place conditions on China's MFN status. So far that has not occurred, but the issue is an ongoing source of irritation in U.S.-China relations, and MFN renewal routinely provokes an emotional debate over a U.S. policy that divorces economic interests from the human rights abuses documented by the State Department and independent groups.

The Clinton Administration has consistently supported MFN status for China and declared the delink of MFN and other bilateral trade issues with China from human rights considerations in June 1994. Repeated efforts have also been made in the U.S. Congress to grant permanent MFN status to China, but despite these efforts, the annual debate continues. On June 3, 1998, President Clinton defied critics in Congress by renewing trade privileges for China [Baker and Dewar 1998]. The debate has taken on additional ardency because of Clinton's trip to China in June 1998.

2.7.3 China's Barriers to U.S. Exports

China maintains several hundred formal non-tariff measures (NTMs) to restrict imports, such as import licensing requirements, import quotas, tendering requirements, and standards and certification requirements. China's restrictive system of trading rights severely limits foreign-invested enterprises' ability to directly import and export, and raises the cost of imported goods. In most transactions, U.S. suppliers are unable to sell directly to their ultimate customer.

In 1992, the United States and China signed a Memorandum of Understanding (MOU) on Market Access that commits China to dismantle most of its trade barriers and gradually open its markets to U.S. exports. In implementing this MOU, China has published numerous previously confidential trade laws and regulations. However, information on China's import quotas has yet to be published on an itemized basis. As a direct result of the Market Access MOU, China has removed over 1,000 quotas and licenses on a wide range of key U.S. exports such as telecommunications digital switching equipment, computers, and medical equipment. As of late 1997, China still retains NTMs on 385 tariff-line items. The final NTM eliminations required under the MOU were scheduled to occur before 1998 [U.S. State Department 1998].

Despite the removal of many quotas and licenses, there have been indications that China is erecting new barriers to restrict imports. Examples include new procedures regarding purchases of large-size medical equipment, and new industrial policies in such areas as automobiles and electronics. China announced plans in early 1996 to begin phasing out import tariff waivers on capital equipment previously available to foreign investors in China.

Despite the Market Access MOU, the imbalance in bilateral trade between U.S. and China continues to widen (see Table 2.12). In order to be acceded into the WTO, China is negotiating with the United States on further reducing tariffs, and has committed to significantly expand trading rights within three years of its accession.

2.7.4 The Issue of WTO Accession by China

China is engaging in intensive negotiations with the United States, the European Union, and Japan concerning the terms under which China will join the World Trade Organization. Western nations are demanding that China implement changes in its trade and trade-related domestic policies to bring them more closely into conformity with the rules embodied in the WTO. The central issue is not whether China becomes a member of the WTO on "commercially viable terms," but rather the length of time it is given for meeting these terms [Lardy 1996].

Protections of intellectual property rights and developed or developing membership status are U.S. concerns regarding China's accession to the WTO. The United States has called on China to improve its protection of intellectual property rights and to expand foreign access to its service market. As a member of the WTO, China will be obligated to abide by the full range of protections offered in the Code on Trade-Related Aspects of Intellectual Property (TRIP).

The WTO helps developing members adjust to global competition by allowing them to maintain existing tariff levels and licensing requirements for an initial three-year period. Thereafter, extensions must be negotiated. Developing countries are also allowed ten years to implement the TRIP code, whereas developed countries must implement the code within one year. China wants greater latitude in meeting WTO requirements in order to protect its electronics, automotive, machinery, chemical, and aviation industries, arguing that its low per capita income makes it a developing country. The United States and other developed countries disagree with this argument, insisting that China is already a major exporter, and that China should therefore enter the WTO under the more demanding standards of a developed country. A possible compromise is under consideration within the Clinton Administration to avoid categorizing China as either a developing or a developed country and instead to spell out its specific obligations [Howell, Nuechterlein, and Hester 1995].

China has taken steps to demonstrate its commitment to trade reforms compatible with WTO policies. Import tariff levels have been reduced significantly since 1992 (see Section 2.7.3). The once dual exchange rate system has been unified (see Section 2.5.4). During the APEC Ministers' Meeting in November 1996, China again pledged to eliminate all non-tariff barriers not consistent with WTO rules, expand foreign companies access to its domestic services market, and allow foreign security firms to establish offices in China. The United States, however, has emphasized that much more progress is necessary and insists that China agree to strict timetables for phasing out discriminatory policies. At the same time, the United States recognizes that considering the large size of China's existing export market and its potential import market, it is in its own interests as well as in international interests to formally include China within the global economic community represented by the WTO.

Although the Chinese Government is committed to joining the WTO, many Chinese officials are concerned about the possible adverse impact of WTO membership on the development of domestic industries. Chinese officials are considering several avenues to protect the Chinese electronics industry after WTO accession, including the use of tariffs, non-tariff trade-protection measures, and exception clauses that pertain to developing countries [Howell, Nuechterlein, and Hester 1995].

2.7.5 U.S. Export Controls

U.S. policymakers employ export controls and economic sanctions not only to address trade and investment disputes, but also to achieve non-economic policy objectives. This has been especially true with respect to China. The level of controls on U.S. exports to China has fluctuated depending on political circumstances in China. During the Cold War period, U.S. export controls on China were very strict. Since November 1983, the United States has imposed export controls on China for products that would make a direct and significant contribution to nuclear weapons and their delivery systems, electronic and submarine warfare, intelligence gathering, power projection, and air superiority.

The Clinton Administration has reduced licensing procedures applicable to a wide range of computer and telecommunications equipment to China. Although there has been steady liberalization of the number of semiconductor devices subject to individual licensing requirements, constraints remain on specific semiconductors and especially on the construction of fabrication facilities. Most advanced semiconductor manufacturing equipment cannot be exported to China without a license from the Commerce Department, applications reviewed on a case-by-case basis. Semiconductor equipment subject to license includes the following [Howell et al. 1995]:

- Metal organic chemical vapor deposition (MOCVD) reactors specially designed for compound semiconductor crystal growth by the chemical reaction between certain materials on the Commerce Control List
- Molecular beam epitaxial growth equipment that uses gas sources
- Lithography equipment using photo-optical or X-ray methods with a light source wavelength shorter than 400 nm or the capability to produce a pattern with a minimum resolvable feature of 0.7 μm or less
- Mask making or semiconductor device processing equipment that uses a deflected focused electron beam, ion beam or "laser" beam with a spot size smaller than 0.2 μm, capability to produce a pattern with a feature size of less than 1 μm, or an overlay accuracy of better than ±0.20 μm (3 sigma)
- Masks or reticles for integrated circuits controlled by 3A01 and for multilayer with a phase shift layer
- Certain "stored program controlled" test equipment

Chapter 3

SCIENCE AND TECHNOLOGY IN CHINA

China's scientific and technological system has made very important contributions to its national economic and social development since the reform began. At the same time, economic reform propels science and technology progress by stimulating innovation and encouraging market competition. In the coming century, China must rely on the advancement of its capabilities in scientific research and technological development to solve an array of problems, such as inefficient industrial structure, technological backwardness, outdated industrial products, insufficient food production, and low labor productivity. To strengthen the science and technology system to an international level and create systematic links between science, technology, and economy has thus become a top priority for Chinese policymakers. This chapter outlines China's science and technology infrastructure, its development status, and the goals of science and technology policy as well as national management of the electronics industry.

3.1 DEVELOPMENT OF SCIENCE AND TECHNOLOGY IN CHINA

China has developed a comprehensive scientific and technological system to support its national economic and social development. This section introduces the relevant research and education organizations in China. Recent developments and foreign cooperation are also discussed.

3.1.1 National Network for Science and Technology Development

As of 1996, China's national network of science and technology research consisted of the Chinese Academy of Sciences (CAS), 5,400 research and development institutions under the supervision of county or lower-level governments, 3,400 research institutions affiliated with schools of higher learning, 13,000 research institutions run by major state enterprises, and 41,000 non-government scientific research-oriented enterprises. In addition, there are more than 160 national academic societies under the jurisdiction of the Chinese Science and Technology Association, all of which have branches in large- and medium-sized cities across the country [IOSC 1997b]. These institutions have formed a national network of research and development that performs basic and applied research and provides technological development and engineering services.

Located in the capital city of Beijing, the Chinese Academy of Sciences is China's highest academic institution and research center of natural and physical sciences. It includes departments of mathematics and physics, chemistry, biology, earth sciences, and technological sciences, a comprehensive studies committee, and an agricultural studies society. In addition, branch academies have been established in provinces, regions, and municipalities where research activities are concentrated.

Operating under the Academy are 123 research institutes employing about 60,000 scientists and engineers. The Academy has selected its 610-member Academic Council, the highest state organ of science and technology consulting, from among scientists, professors, and engineers who have made significant contributions to their fields [Qin 1997]. The Chinese Academy of Engineering, established in 1994, is the country's highest honorary consulting and academic organization in China's engineering community.

3.1.2 Status of China's Science and Technology Development[1]

China's total national expenditure on research and development in 1995 was $2.65 billion, a growth rate of 13.8 percent over the previous year. It accounted for only 0.5 percent of China's gross domestic product (GDP), compared with 2.4 percent in the United States in the same period. Of the total expenditure on research and development at the national level, 6.1 percent was spent on basic research, 39.8 percent on applied research, and 54.1 percent on technology development. In the same period, governmental research institutions spent (12 billion yuan or 44% of the total); enterprises spent $1.1 billion (9.1 billion yuan or 31.9% of the total); university research institutions spent $470 million (3.9 billion yuan or 13.7% of the total); and other research entities spent $350 million (2.9 billion yuan or 10.4% of the total national expenditure). Total national expenditures in research reached $10.7 billion in 1995, of which $3.6 billion was appropriated by the PRC government. The central government appropriated 71.3 percent and local governments appropriated 28.7 percent.

In 1995, there were 1.4 million scientists and engineers engaged in science and technology in China. Among them, 422,000 were involved in research and development—6.8 researchers per 10,000 population. By comparison, in 1993 the ratio was 74.3 per 10,000 in the United States. The total number of Chinese technicians, scientists, and engineers reached 2.58 million in 1995. Among them, 665,000 (25.8%) were involved in research and development, 1,316,000 (51%) were involved in enterprises, and 600,000 (23.2%) were involved in education.

In 1995, the China Patent Office received 83,045 applications and granted 45,064 of them. Domestic applicants submitted 68,880 and foreign

---

[1] The information in this section is from the *Science and Technology Statistics Databook (1996)*, compiled by China's State Science and Technology Commission.

applicants submitted 14,165. Patents granted to domestic applicants numbered 41,248 and patents granted to foreign applicants numbered 3,816. In the same year, domestic science journals published 107,991 articles, an increase over 107,492 in 1994. In 1995, there were approximately 1,200 journals published in China. The combined number of articles catalogued by SCI, ISTP, and EI indexes reached 26,395 in 1995 and 24,584 in 1994.

The export value of high-tech products reached $10 billion in 1995. It represented a 59 percent increase over the previous year and accounted for 6.8 percent of China's total exports in that year. The value of imported high-tech products reached $21.8 billion, 6 percent higher than 1994, and accounted for 16.5 percent of total import value in 1995. International trade data for certain technology categories reveals that China's top priority is development of its information technology infrastructure. The leading exports were computers and telecommunications ($6 billion), electronics ($1.3 billion), and life sciences products ($990 million). The leading imports were in computers and telecommunications ($8.3 billion), computer-integrated manufacturing equipment ($6.3 billion), aerospace and aviation ($1.5 billion).

3.1.3 Science and Technology Education at the University Level[2]

Universities in China serve a small but rapidly growing percentage of the population (see Chapter 1). In 1996, of the 1,032 ordinary colleges and universities in China (including colleges for professional training), 419 had the ability to train graduate students. These had some 140 doctoral programs and 470 master's programs in the fields of networking, communications, computer and computer applications, software, wireless electronics, microelectronics, and electronic components and devices; every year they confer about 200 PhD degrees, 2,500 master's degrees, and 40,000 bachelor's in those fields. So far there are 101 national laboratories: among them, 30 serve the fields of electronic information and related subjects.

The higher-level institutions offering education in electronic science and technology can be divided into four classes:

1. Institutions directly under the control of the State Education Commission: Peking University, Tsinghua University, Fudan University, Shanghai Jiaotong University, Zhejiang University, and the Southeastern University.

2. Institutions under the administration of specific government agencies, such as the University of Space Flight and Aviation and Harbin Industrial University of the Department of Space-flight, and the University of Posts and Telecommunications of the Ministry of the Posts and Telecommunication (MPT).

---

[2] Statistics in this section were provided by China's State Planning Commission, 1997.

3. Regional colleges and universities, such as Beijing Industrial University and Beijing University of Chemical Technology.

4. Institutions under the administration of the Ministry of the Electronics Industry (MEI): the University of the Science and Technology of Electronics, Xian University of the Electronic Industry, Hangzhou Institute of the Electronic Industry, Jilin Institute of the Electronic Industry, and Beijing Institute of Information Engineering.

3.1.4 The National Centers of Engineering Research

The State Planning Commission started to actively promote establishment of national centers of engineering research at the beginning of the seventh Five-year Plan (FYP) (1986-1990). The main purposes of the centers are to convert significant scientific research results into useful, economically viable products; to solve engineering problems related to key areas of industrial development; and to explore ways to integrate science and technology into the economy. The National Centers of Engineering are being developed as limited liability corporations focusing on the following areas:

*   systems integration engineering,
*   process engineering, and
*   training of a new generation of first-rate engineers and technicians capable of innovation, development, and interdisciplinary cooperation.

By the end of 1996, 74 national engineering research centers had been established in China, of which 34 are dedicated to the fields of electronics and information technology, with a total investment of more than $2 billion (17 billion yuan). The main scope of these centers includes microelectronics, computers and software, communications, automation, and electronic product and process development. Table 3.1 lists the names of these 34 centers and their locations.

3.1.5 Research Institutes Related to the Electronics Industry

Five types of research institutes in China are related to the electronics industry:

1. Forty-eight institutes under the supervision of the MEI focus on research subjects such as technology, audio and video, standards, reliability and environmental testing, computer, communication, optical fiber and cable, applied magnetics, applied optics, vacuum electronics, semiconductors, electric cable, microelectric motors, electronic equipment, lasers, electronic measurement, electronic materials, and microwave technologies.

2. Institutes under the supervision of the Chinese Academy of Science (CAS) include the CAS Semiconductor Institute, CAS Shanghai Metallurgical Institute, and the CAS Microelectronics Research Center.

Table 3.1
National Electronic and Communication Engineering Research Centers

Main Research Field	Location
Application of Electric Power Electronics	Hangzhou
Automation of Industrial Course	Shanghai
CAE for Rubber & Plastic Mould	Zhenzhou
Communication Software & ASIC Design	Shijiazhuang
Computer Software	Shenyang
Computer Software for Petrol & Gas Exploration	Beijing
Digital Wireless Telephone System	Xian
Digitized Product for Video & Audio	Nanjing
Electric Converting Technique	Zhuzou
Electric Power Electronics	Xian
Electric System Automation	Nanjing
Electric Transmission	Tianjing
Electric Transmission & Energy Saving Technique	Beijing
High Grade Micro-Controlling	Shenyang
Industrial Automation	Hangzhou
Laser Fabrication	Wuhan
LSI CAD	Beijing
Manufacturing Industry Automation	Beijing
Microelectronics	Beijing, Shanghai
Mobile Communication	Guangzhou
Mould CAD	Shanghai
New Technique for Electronic Press	Beijing
New Type Electric Source	Tianjing
Optical Disk & Application	Shanghai
Optical Disk System & Application Technique	Beijing
Optical Fiber Communication Technique	Wuhan
Optoelectronic Devices	Beijing
Robot	Shenyang
Semiconductor Material	Beijing
Sensor	Shenyang
Ship Design Technique	Shanghai
Ship Transportation Control System	Shanghai
Software Engineering	Beijing, Guangzhou
Telecom. Switches & Software Supporting System	Xian

Source: State Science and Technology Commission, 1997.

3. University institutes include the Microelectronics Institute at Peking University, the Microelectronics Institute at Tsinghua University, the Materials Science Institute at Fudan University, the LSI Institute at Shanghai Jiaotong University, and the CAD Institute at Hangzhou Electronic Industry College.

4. Other institutes directly under various ministries include the Microelectronics Institute of the Space Flight Industry and the Communications Institute of the Ministry of Posts and Telecommunications.

5. Institutes within enterprise itself include the Central Institute of the Huajin Group Company.

In 1995 there were 63,586 employees in China's 48 S&T institutes directly under MEI; among them were 35,025 professionals (1,864 with graduate degrees, 18,166 with undergraduate degrees), comprising 55 percent of total employees. In 1995 the total income of those 48 institutes was $482 million (4 billion yuan).

3.1.6 Sino-U.S. Science and Technology Cooperation

Since 1978, China has sent over 280,000 people to study abroad; about 90,000 of them had returned to China as of 1997. The long-term brain drain from Asia to the United States has begun to reverse itself. The Four Dragons (Hong Kong, Singapore, South Korea, and Taiwan) are now able to offer attractive working conditions and salaries to new graduates and experienced high-tech professionals. However, China still has a long way to go in improving working conditions to attract more people back home.

Apart from sending people to other countries to study, China's industrial and scientific community has been conducting academic discussions and technology exchanges with its counterparts in foreign countries. Since 1986, China has herself held many successful international conferences on microelectronics technology, ASIC techniques, integrated circuit (IC) materials, characterization techniques for IC materials and processes, computer technology, and so forth. Every year, China also sends people to attend international conferences on information technology to present the latest results of Chinese research and to learn about technical trends elsewhere. In addition, Chinese businessmen and scientists help to strengthen domestic and international communication and cooperation by co-writing books, visiting other organizations and enterprises, and so forth. Now China is establishing cooperative relationships with foreign scholars through global electronics and information networks.

China and the United States concluded a bilateral science and technology cooperation accord in January 1979 when Deng Xiaoping made a visit to the U.S. A Sino-U.S. Agreement of Cooperation in Science and Technology was signed between China and the United States. Since then, scientists and engineers in the two countries have carried out fruitful cooperation in numerous fields with the guidance of the agreement and a series of other protocols and memorandums signed afterwards by concerned authorities in both countries. In 1995, more than 20,000 people in the science and technology community conducted exchange activities over 5,700 projects [*China Science and Technology Newsletter* April 1997].

International science and technology collaboration is one of the most effective ways to popularize modern civilization, promote social progress, and strengthen economic cooperation. Both Chinese and American governments have attached great importance and provided strong support to the Sino-U.S. science and technology cooperation.

3.2 SCIENCE AND TECHNOLOGY INFORMATION NETWORKS

As China becomes more and more open to the outside world, the construction of a national information infrastructure connecting to the Internet to facilitate trade and information exchange has become a top priority for national economic and social development. China must develop and utilize science and technology information networks to boost the productivity of manufacturing, financial, and other sectors. As of 1998, four major nationwide Internet-based networks in China connect with the international Internet backbone: *CHINANET, China Golden Bridge Network (CHINAGBN), China Education and Research Network (CERNET)*, and *China Science and Technology Network (CSTNET)*. *CHINANET* and *CHINAGBN* are primarily for commercial usage, while *CERNET* and *CSTNET* are dedicated to academic research only.

3.2.1 Public Commercial Networks

CHINANET was started in 1994 by the Ministry of Posts and Telecommunications (MPT) to provide various Internet services to public users and to promote the commercialization of the information network. The structure of the *CHINANET* includes eight regional network centers that cover thirty-one provinces and cities. As of 1997, there were more than two hundred business subscribers and sixty thousand individual subscribers, with more cities working on expanding access to the networks [AATS 1997].

CHINAGBN is a nationwide public economic information processing network sponsored by the Ministry of the Electronics Industry (MEI). The network was first implemented in March 1993, and is one of the most important components of China's Golden Bridge Project. This Golden Bridge project is being implemented to form an information network for the national public economy, connecting all levels of government and middle and large enterprises. The major objectives of the *CHINAGBN* are to establish a public economic information network to interconnect the heterogeneous private network of multiple departments and sectors, and to establish a computer information system for government agencies and private enterprises. As of 1997, the network was set up in twenty provinces and cities in central and eastern China, especially the coastal cities, and is able to transmit data information, picture images, and voice through the satellite network.

3.2.2 Academic Research Networks

The development of China's academic research network can be divided into two main stages. The first stage was the informal connection stage. During this period, data transmission speed was slow and connection methods were limited. Users often used long distance telephone lines to access online databases in other countries. In March 1993, the Institute of High-Energy Physics (IHEP) of the Chinese Academy of Sciences (CAS) opened a 64kbps international data channel that was also connected with the Stanford Linear Accelerator Center in the United States. As a result, users were able to utilize dial-up lines and public data networks to exchange information directly.

The second stage was the formal introduction of the Internet into China in April 1994 by the CAS's Computer Network Information Center (CNIC) in Beijing. Established in 1990, CNIC oversaw the construction of a supercomputer center in the Beijing area, and facilitated the use of supercomputers by building networks that linked together more than thirty research institutes with fiber optic cables. The networking was completed and the 64 kbps line was linked to the Internet in 1994.

China's academic research network experienced four different project development stages: the Networking and Computing Facility of China (NCFC), Chinese Academy of Sciences Network China (CASNET), Education and Research Network (CERNET), and China Science and Technology Network (CSTNET).

3.2.2.1 Networking and Computing Facility of China

The Networking and Computing Facility of China (NCFC) was jointly invested in by the State Planning Commission and the World Bank's Key Scientific Development Project. The facility is run by the Chinese Academy of Sciences' Computer Network Information Center (CAS-CNIC) and linked together with Peking University and Tsinghua University. The NCFC network is made up of four parts: Wide Area Network (WAN), Urban Area Network (main network), Campus Network, and Local Area Network (LAN).

NCFC's services include domain name service,[3] a mail server, an FTP server, a Gopher server, a news server, and a WWW server. The CAS-CNIC also established network supervisor facilities, which compile statistics about international communication volume [*China Electronic News*, October 11, 1997].

---

[3] The first level of China's domain name system is .cn, with two types of second level domains. The first type is vertical domain names, such as .edu, and .net. The second type is the thirty-three horizontal domain names categorized by cities, provinces, and regions, for example .bj (Beijing), .sh (Shanghai), .tj (Tianjin), and .zj (Zhejiang).

3.2.2.2 Chinese Academy of Sciences Network

The Chinese Academy of Sciences Network (CASNET) is CAS's national research network. CASNET is divided into two parts. The first part is the branch campus area network established in 1992. In addition to the main campus and fifty or so research institutes in Beijing, CAS has twelve branch campuses distributed in cities around the country.

The second part of the CASNET is using long distance channels to connect all branch schools and research institutes to the NCFC network center's WAN. At the end of 1995, CASNET successfully connected the twelve branch campuses and various research institutes to the WAN in Beijing.

3.2.2.3 China Education and Research Network

The Chinese Education and Research Network (CERNET) is the first nationwide education and research computer network in China[4]. The CERNET project is funded by the State Planning Commission and managed by the Chinese State Education Commission. The main objective of the CERNET project is to establish a nationwide network infrastructure to support education and research in and among universities, institutes, and schools in China using up-to-date telecommunication and computer technologies.

CERNET has adopted a three-level management hierarchy—nationwide backbone, regional networks, and campus networks. The CERNET national backbone uses the Digital Data Network (DDN) offered by China's Ministry of Posts and Telecommunication. The CERNET has one national network center (located at Tsinghua University) and ten regional network nodes distributed among ten universities covering most of China's geographical area. As of 1997, there were 1,075 universities, 390,000 professors and staff members, 94,200 graduate students, and 2,184,000 undergraduate students connected to the system [China National Network Center, 1998].

CERNET is connected to the international network system and is on a par with the most advanced international systems, based on the advanced SMNP network system. Establishing CERNET has spurred the development of various regional information-sharing networks. The cities of Shanghai and Gongzhou and Jiangsu Province have already set up education networks of their own, and Tianjing City, Jilin Province, and Liaoning Province have begun to plan their own information networks.

3.2.2.4 China Science and Technology Network

The China Science and Technology Network (CSTNET), which is based on CASNET and NCFC, is a network that links together the various Chinese science and technology personnel, management departments (e.g., National Natural Science Fund Committee), and relevant government departments (e.g., the National Patent Bureau) outside of the CAS system. Construction

---

[4] CERNET's Internet website is http://www.net.edu.cn

started in 1990, and the basic network was completed in 1992. CSTNET connected more than thirty research institutions in the Beijing area, including Peking University and Tsinghua University. Following the success of the NCFC, CSTNET continued to work on connecting institutions outside of Beijing under the name of the Hundred Research Institutions Connection Project. The project was concluded with success in 1995 [AATS 1997].

3.3   SCIENCE AND TECHNOLOGY POLICY

National policy set the general direction for resource allocation in Chinese science and technology development. This section presents the historical background of China's science and technology policy in the post-reform era, goals and tasks for national science and technology development for the twenty-first century, and national management of the electronics industry.

3.3.1   Historical Background

The year 1982 was significant in China for science and education policy in general. In that year the Central Committee of the Chinese Communist Party and the State Council, which set national policy in economic as well as other matters, drafted a strategic policy that tightly linked the development of science and technology with that of the economy as a whole. The State Planning Commission and the State Commission of Science and Technology, along with concerned departments of the State Council, were charged with designing national plans to tackle the key problems of science and technology.

The primary organization in charge of this campaign has been the State Planning Commission, which handles organization, coordination, and administration of various tasks such as making plans, determining and verifying projects, signing contracts, allocating funds, and approving projects.

Various state departments and local planning committees are in charge of implementing the key projects in their areas of concern. Participating enterprises and institutions must complete the tasks of the campaign in accordance with campaign contracts. Also taking part in the campaign's administration are the Chinese Academy of Science and the State Education Committee. The Division of Science and Technology of the Ministry of the Electronic Industry administers the electronics-related programs of the campaign.

In March 1986, China launched a high-tech development program known as Project 863, which called for organizing scientists and engineers to keep up with the world's latest achievements in seven high-technology fields: biotechnology, space technology, information technology, laser technology, automation technology, energy technology, and development of new materials. The goal was to stimulate thought so that breakthroughs could be made to narrow the gap between China and the developed world in scientific development.

Since 1988, China has been carrying out the Torch Program—the State Science and Technology Commission effort to promote the commercial application of research output and facilitate the development of advanced technologies. As of 1997, 120 high-technology development zones had been set up around the country [Qin 1997]. One facet of the Torch Program has been the promotion of science parks with the expectation that the parks will breed innovation. However, the result so far has not been satisfactory due to insufficient competition in the market to drive innovation.

In order to effectively promote the development and practical application of high technology, the Chinese government promulgated the *National Compendium on Intermediate- and Long-Term Scientific and Technological Development* in March 1992. This compendium laid out a grand blueprint highlighting the essential foci of China's plan to develop new advanced technologies and the nation's strategic objectives toward the year 2020 [Qin 1997].

3.3.2 Science and Technology Policy for the Twenty-First Century

In May 1995, the Central Committee of the Chinese Communist Party (CPC) and the State Council issued the *Decision on Accelerating Scientific and Technological Development*, which outlines China's development strategy for the next several decades. It decrees that science and technology research must become closely tied to the market, and institutions of higher education should seek to form joint ventures with Chinese or foreign venture capitalists in order to accelerate the transfer of science and technology to China. The major goals set up by the State Council include (1) boosting China's annual science and technology research spending to 1.5 percent of GNP by the year 2000; (2) protecting the environment and achieving sustainable development; (3) achieving indigenization of science and technology creation capabilities in key areas of manufacturing technology and systems design. Specifically, China should set out to achieve by the year 2010 the following goals [State Council 1995]:

- strengthen the science and technology system and create organic links between it and the economy;
- make science and technology prosperous and train a new generation of knowledge workers;
- achieve significant gains in agricultural and industrial R&D, as well as basic research and high technology;
- increase the proportion of economic growth attributable to S&T progress;
- match the approach to science and technology of the advanced countries at least in some fields;

- considerably increase China's capability to create technology indigenously and master key industrial and systems design technologies.

The principles guiding China's science and technology policy in the coming century should include the following specifics [State Council 1995]:

- the prime mission is to solve important questions of social and economic development. Goals, organization structure, regulations, and plans should reflect the organic links between science and technology and the economy.
- reform propels science and technology progress. The government should make full use of market forces to stimulate progress.
- insist on the combination of autonomous research and the importation of foreign technology to enhance the transformation of S&T progress into results that can be used by industry.
- bearing in mind international science and technology trends and China's situation, Chinese science and technology capacity should focus on limited objectives and concentrate its resources where suitable for achieving breakthroughs.
- respect knowledge and talent and permit people to develop them. In academic research insist on academic freedom, and make important decisions scientifically and democratically.

The State Council's report emphasized the need for China to use market forces to propel indigenous technologies. Realizing that China cannot compete in all areas, the report calls for concentration in a few technologies and on boosting the role of high technology in delivering competitive products. The report also focuses on boosting agricultural and industrial production as well as for the importance of information networks. In addition, the State Council calls for management changes. There is a need for consolidating research institutions, increasing the mobility of personnel between organizations, improving the flow of information, encouraging competition and open bidding on projects, linking pay to economic results, protecting intellectual property, and allowing talent to flourish through academic democracy. The report is unclear on how research will be funded. However, it does focus on the need for the market to support applied research and discusses for the first time the role of venture capital in funding research and development.

3.3.3 Main Task for Science and Technology Development Under the National Ninth Five-year Plan

Under China's Ninth Five-year Plan (FYP) for national economic and social development (1996-2000) and the *Long-Term Targets Through the Year 2010* approved by the fourth session of the Eighth National People's Congress held in March 1996, science and technology work is obliged to serve the needs of the country's economic and social development and to help improve labor productivity, enhance the overall strength of the nation, and raise people's living standards. The specific tasks are [IOSC 1997]:

- stepping up technological and product development according to market demand, accelerating commercialization of research output, and pooling resources to develop technologies of importance to China's economic and social development;
- making positive efforts to develop technologies and high-tech industries, with a focus on electronics, information technology, bioengineering, new materials, alternative energy sources, aeronautics, astronautics, and oceanographic industries, while striving to approach or meet the international standards in some of these areas;
- strengthening basic research, following the latest developments in global research and development, and striving for major breakthroughs in areas where China has advantages.

During the fifteenth National Congress of the Communist Party in September 1997, President Jiang Zemin emphasized that government should deepen the reform of the management systems of science and technology education to promote their integration with the economy [Jiang 1997]. In addition, research institutes and institutions of higher education should combine production, teaching, and research by entering into cooperation with private enterprises to solve the problems of segmentation and dispersal of strength in science, technology, and education.   Finally, transferring intellectual resources from overseas and encouraging those studying abroad to return and render their service to the homeland must be actively supported [Jiang 1997].

3.3.4   National Management of the Electronics Industry

In October 1949, the government established a national bureau to oversee the development of telecommunications capability within China—the Telecommunications Industry Bureau. The various bureaus having to do with the electronics and telecommunications industries have undergone several metamorphoses since 1949. The most important of these occurred in 1982 when the Ministry of the Machinery Industry merged with the National General Bureau of Telecommunications, Broadcasting, and Television and with the National General Bureau of Electronics and Computers to become the Ministry of the Electronics Industry (MEI).[5] Although MEI is the ministry charged with overseeing the development of China's electronics and telecommunications industries, some electronics enterprises and research institutes are still under the jurisdiction of other ministries, such as the

---

[5] The head of that Ministry from 1983-85 was Jiang Zemin, who has been China's President since March 1993.

Ministries of Posts and Telecommunications (MPT); Aerospace and Aviation; Railways; Metallurgy; Broadcast, Film, and Television; or Commerce; still others are under the jurisdiction of the Chinese Academy, China Nonferrous Metal Industrial Corporation, or the Bureau of the Construction Materials Industry.

3.3.4.1 Ministry of Electronics Industry

China's Ministry of Electronics Industry (MEI) is an administrative department of the State Council in charge of the national electronics industry. MEI is responsible for the administration of the manufacturing of electronics systems, equipment, and products; the research and development of electronic technologies; the development of computer software; and the application of information technology as well. The main tasks of MEI include formulating national strategy, guiding principles, administering regulations, as well as maintaining technical and quality standards for China's electronics industry. Although some enterprises, research institutes, and universities are under MEI's direct leadership, most local enterprises are administered by the electronics bureaus of their provinces or cities. There are several other government ministries and bureaus whose projects and authority overlap that of MEI.

The organizational setup of the MEI includes the minister and several vice-ministers. Under them are a General Office and ten other departments. The General Office is a comprehensive administrative office in charge of the coordination of the ministry's general affairs. The major responsibilities of the ten departments under MEI are as follows [MEI 1997]:

- *Overall Planning*—responsible for the programming and planning of the electronics industry;
- *Economic Operation and Structure Reform*—responsible for establishing targets for production, managing information and statistical work, coordinating efforts by relevant departments, and organizing and coordinating international bidding;
- *Technology and Quality Supervision*—responsible for devising programs and plans for science and technology development, promoting the commercialization of electronics technologies, supervising the protection of intellectual property rights, setting up technical and quality standards, and instituting and supervising quality management programs;
- *Department of Telecommunication Products and Systems*—responsible for putting forward policies on the development of major electronics systems and equipment, establishing plans for building China's manufacturing capabilities, and organizing and coordinating the development of major electronics and engineering systems;
- *Computer and Information Technology Pervasiveness*—responsible for managing the computer industry and coordinating the development of electronics information services;

- *Major Projects of Components and Devices*—responsible for implementing key projects for microelectronics and other basic products, and for organizing and coordinating the application and manufacturing of microelectronics products;
- *Economic Regulation and State Asset Supervision*—responsible for taking part in the formulation of various state economic policies, managing research funds, and managing state properties and enterprises under MEI control;
- *International Cooperation*—responsible for foreign economic and technical cooperation and exchanges in the electronics industry, overseeing the import and export of electronic products, taking charge of the foreign-related activities in projects involving overseas investment, and handling foreign affairs for the industry;
- *Department of Special Purpose Equipment*—responsible for organizing and coordinating the research, development, and production of electronics products, equipment, and systems for the national defense;
- *Department of Personnel and Education*—responsible for making policies for personnel and worker education, managing study-abroad programs, and managing the universities directly under MEI supervision.

3.3.4.2 Ministry of Posts and Telecommunications

The Ministry of Posts and Telecommunications (MPT) is a functional department under the State Council. MPT is both the state regulator and operator. It is responsible for the macro-control of the nation's posts and telecommunications industry, making overall plans, coordinating projects, and supervising operations. Telecommunications operations are now handled by an MPT unit called the Directorate Generale of Telecommunications, also known as China Telecom. MPT exercises centralized control of the nation's public communications networks and the communications market, maintains the communications order, and ensures the integrity, uniformity, and advanced features of the nation's public communications networks.

3.3.4.3 The New Ministry of Information

A session of the Chinese State Council passed a resolution during the National People's Congress in early 1998 to restructure and downsize the government by combining various government ministries and to cut the total number of employees in half. Under the State Council's plan, the old Ministry of Electronics Industry, the Ministry of Posts and Telecommunications, and segments of other ministries have been combined into a new Ministry of the Information Industry (MII). In addition, parts of the Ministry of Film, Radio, and TV, as well as the Ministry of Aerospace and Aeronautics (those functions responsible for network system management), will also be moved into the MII.

The consolidation of MEI and MPT into one MII seems to be logical given the convergence of the electronics and telecommunications technologies market worldwide. However, this move to consolidate and downsize China's inefficient and over-staffed government ministries is going less than smoothly, and has raised concerns in the nation's electronics and information technology industries because the two old ministries that were to be replaced are still operating, as is the new ministry intended to replace them [Carroll 1998c]. It is still not clear about the exact time frame and outcome of this consolidation.

Chapter 4

THE DEVELOPMENT OF CHINA'S ELECTRONICS INDUSTRY

During the late 1800s and early 1900s, China's investments in national telegraph and telephone networks and in electrical and electronics technologies were heavily subsidized by foreign companies such as General Electric (U.S.), Nippon Electrical Company (Japan), and A.R. St. Louis (Canada). The early twentieth-century electronics factories in China conducted mainly simple assembly and equipment maintenance.

In the 1930s and 1940s, China's electronics enterprises reflected the political divisions within the country between the two factions vying for power: the Nationalist Government, led by the Kuomintang (KMT) party, and its ideological rival, the Chinese Communist Party (CCP). The KMT, with Western support, built factories that produced simple electrical equipment and devices such as light bulbs, vacuum tubes, radio parts, motors, wires, batteries, telephones, and switchboards. In 1946 there were about two hundred factories and seven thousand workers engaged in this work; the gross output value amounted to only 3.6 million yuan, or only 0.02 percent of the gross output value of all national industries.[1] In the same period, the CCP maintained a simple radio network of its own and operated small-scale factories that produced radio devices and parts to support its military needs. As early as 1934, the CCP had established a commission to oversee military telecommunications work.

The devastation of World War II and civil war, the poverty of the country, natural disasters, the failure of the "Great Leap Forward" campaigns of the 1950s, and the chaos brought about by the "Cultural Revolution" in the 1960s and early 1970s greatly impeded development of China's industrial sectors. Still, an overview of the country's national bureaucracy and plans in support of electronics development indicate the importance that the Chinese Communist government has consistently placed on this sector of its economy, even though it was not always possible to make the rapid progress that was desired.

In general, China was a late and slow starter in developing its electronics and telecommunications industries. However, significant progress has been

---

[1] Data was provided by the State Science and Technology Commission, 1997.

made in recent years through the implementation of a series of national development plans and government programs. Recent achievements in the electronics field demonstrate China's determination to become a world-class player. This chapter summarizes the development of China's electronics industry. Foreign trade and cooperation as well as national planning in China's electronics industry are also discussed.

4.1   THE ELECTRONICS INDUSTRY IN CHINA'S NATIONAL FIVE-YEAR DEVELOPMENT PLANS

As noted in Chapters 1 and 2, China's national economic goal-setting and planning mechanism is the "Five-Year Plan" (FYP). Development of the country's electronics capabilities has hinged on its ability to define and achieve reasonable goals in the FYPs. The following sections summarize the technological and industrial development of China's electronics industry since the first five-year plan.

4.1.1   The First Five-Year Plan, 1953-1957
The twin electronics goals of the first FYP were to modernize and strengthen the radio and communications aspects of national defense and to set up automatic telephone switchboard factories for the civilian network. Soviet assistance was promised for establishing 156 key national projects, of which nine were related to electronics, such as building the Beijing Electronic Tube factory.

During this first FYP China laid the foundation for development of its electronics industry through establishment of scientific research and educational institutions and through long-range planning, as detailed in the 1956 "Long-Range Plan for the Development of Science and Technology from 1956 to 1967." This plan listed key electronics projects of national importance: telecommunications and broadcasting systems, research and development of radio electronics, semiconductor technology, and computer and radio technology for national defense.

4.1.2   The Second Five-Year Plan, 1958-1965
The goals of China's second FYP electronics industrial development were in the following three major defense-oriented areas:

1. development and construction of electronics equipment for a ballistic missile capability.
2. development and construction of electronics equipment to support atomic energy and aviation development.
3. establishment of factories to produce electronic measuring instruments and specialized electronic equipment.

The Soviet Union's technical and financial assistance was a key ingredient of the plans of this period. Initially, there was significant progress building new factories and institutes focused on electronics technologies. Research and development institutes built with Soviet assistance included the Chendu Radio Communication Institute, Beijing Radio Components and Materials Institute, Beijing Electronic Tube and Transistor Institute, and Nanjing Radar Institute.

By 1959, twenty-five Chinese national universities were teaching classes in electronics fields, including radio technology, vacuum electronics, semiconductor technology, and computer science. Leading universities in these fields were Tsinghua, Fudan, Jiaotong, and Zhejiang Universities.

Early in this period, the national "Great Leap Forward" campaign commenced. Due to political differences, the Soviet Union abandoned its commitment to assist in China's economic development. Development of China's electronics industry stalled in the period 1960 to 1961. As a result, factory efficiency and profit margins took a nosedive. It was not until 1965 that the production quotas in electronics products were restored to the level of early 1960.

4.1.3   The Third and Fourth Five-Year Plans, 1966-1975

China's goals for the electronics industry in the third and fourth FYPs slightly broadened to speed up construction of the electronics industry to meet the national strategic and defense needs (e.g., for radar and aviation equipment) and to devote major effort to developing basic products to help build the national economy.

Both goals had the greater aim of catching up with and exceeding the technological ability of advanced countries in certain key areas within five years. However, the excesses of the Cultural Revolution deflected economic development throughout China. Cut off from the world and racked by civil violence and political extremism, China fell further behind in its industrial development.

4.1.4   The Fifth and Sixth Five-Year Plans, 1976-1985

The Cultural Revolution formally ended in October 1976. As the national economy began to recover, the electronics industry took on a new look during Fifth and Sixth FYPs. China revised its military and civil electronics production structures so those military factories produced both military and civilian products. Another adjustment increased the ratio of consumer products to capital equipment in total output of electronics products.

During this period, the Ministry of the Electronics Industry was established. In addition, a so-called "lead group" focused on the electronics industry was formed within the State Council. This lead group was first constituted as the "Electronic Computer and LSI Lead Group" in 1982, with then Vice Premier Wan Li as the head of the group. Jiang Zemin, who was then Vice Minister of the Electronics Industry, was a member of this group. The group was renamed to the "Lead Group to Promote Electronics" in 1984, with Vice Premier Li Peng as its head.

4.1.5   The Seventh Five-Year Plan, 1986-1990

During the Seventh FYP, the production level in the electronics industry began to accelerate, despite the instability of market prices and uneven economic growth in regions. The electronics industry's annual target rate of growth was 16 percent for the Seventh FYP, but the actual growth rates fluctuated widely from 1986 to 1990.[2] In 1988, the gross output value of the electronics industry fell just short of 1990's goal of 60 billion yuan (~$16.1 billion). The 1990 gross output value was 67.3 billion yuan (~$14.1 billion), which was 2.4 times that of 1985.

Exports of electronics goods became increasingly important to the Chinese economy during this period. According to Chinese customs statistics, China's total exports of electronics products during the Seventh FYP were valued at $10.5 billion; the average annual growth rate of export was about 66 percent.

The Seventh FYP was the first in which projects of the campaign to tackle key science and technology problems were included. In this period there were five large projects relating to electronics: VLSI technology, computer systems, computer software, communication technology, and electronic materials. Major achievements of this period included development of a number of indigenous electronic devices and systems, including a GaAs integrated circuit, a 700 keV high-energy ion planting machine, the Taiji 2000 series super microcomputer, the Huasheng 4000 series workstation and 0500 series 32-bit high-performance computer, a shipborne microwave measuring and control system, and an air transportation control system.

4.1.6   The Eighth Five-Year Plan, 1991-1995

During the Eighth FYP, China's electronics and information technology industry continued its rapid development. The industry's key economic goals, which were set at the beginning of the plan, were accomplished two years ahead of the 5-year mark. In 1995, the gross value of industrial output reached $29.4 billion (245.7 billion yuan). Exports of electronics products were valued at $16.5 billion, which for the first time surpassed the value of electronics imports ($16.1 billion) [*China Electronics Industry Yearbook 1997*, 234].

---

[2] The actual annual growth rates of China's electronics industry from 1986 to 1990 were -29%, 143%, 40%, 5.8%, and 6%, respectively. Statistics were provided by the State Science and Technology Commission, 1997.

During this period, total production of color televisions, radios, audio cassette recorders and some other electronic components ranked first in the world as China built up its mass production capabilities and improved the global competitiveness of its products.

During this FYP, the structure of China's electronics industry became increasingly sophisticated. A number of big electronics/information technology companies such as Changhong, Caihong, Shanghai Broadcasting and Television, Panda, Hualu, and Lianxiang came into being. Enhancement of scientific research and technological development programs resulted in significant progress, such as development of VLSI devices, the Panda ICCAD system, and an erasable and recordable CD. Some of China's mainframe and microcomputers began to achieve international technology levels, and Chinese system software and platforms began to exhibit characteristics tailored to the domestic market. Chinese-made large local digital switchers, such as the 04,601,08 series, have entered mass production. In addition, three "Golden Projects" were organized that opened new domestic markets for the electronics and telecommunications industries.

In the Eighth FYP, 17 major electronics projects were organized under a campaign to tackle key problems in science and technology. There were more than four hundred subprojects in fields such as VLSI microfabrication technology, mass production of 1-1.5 μm VLSI technology, CAD/CAM, microfabrication tools, microelectronics materials, microanalysis techniques, advanced personal computers and workstations, computer software, electronic devices for electric power, HDTV techniques, and equipment for air traffic control. The central government allocated 555 million yuan (~$110 million) to support these key electronics projects within this FYP. Some eight thousand scientists and technicians participated in these projects.

4.1.7 Evolution of China's Electronics Industry Through 1995

The steady growth of China's electronics and telecommunications industries in the first eight five-year plans is shown in the following two tables. Table 4.1 shows total national output and national investment in the infrastructure of the electronics industry, as well as number of employees.

Table 4.2 lists key electronics products and systems developed within China and the organizations that developed them.

Table 4.1 Output Value, Number of Employees, and Infrastructure Investment

Five Year Plan	Year	Output Value (million yuan)	Employees (thousands)	Investment (million yuan)
First FYP	1953	42	16	555
	1954	45	21	
	1955	46	35	
	1956	75	79	
	1957	107	90	
Second FYP	1958	323	167	935
	1959	579	201	
	1960	751	283	
	1961	251	257	
	1962	217	228	
	1963	259	230	
	1964	354	253	
	1965	570	293	
Third FYP	1966	832	334	2,115
	1967	670	363	
	1968	660	402	
	1969	1,180	482	
	1970	2,230	650	
Fourth FYP	1971	3,083	804	
	1972	3,122	844	
	1973	2,966	853	
	1974	3,361	867	
	1975	5,079	922	
Fifth FYP	1976	5,474	1,020	3,811
	1977	6,950	1,070	
	1978	7,644	1,160	
	1979	8,197	1,240	
	1980	10,060	1,330	
Sixth FYP	1981	10,859	1,360	
	1982	11,010	1,350	
	1983	14,318	1,380	
	1984	21,447	1,420	
	1985	28,636	1,580	
Seventh FYP	1986	30,020	1,570	6,940
	1987	42,740	1,560	
	1988	59,920	1,610	
	1989	63,420	1,650	
	1990	67,270	1,680	
Eighth FYP	1991	88,630	1,730	6,400
	1992	108,700	1,710	
	1993	139,600	1,680	
	1994	186,200	1,670	
	1995	245,700	1,700	
Ninth FYP	1996	298,200	1,710	

Source: State Science and Technology Commission, 199

Table 4.2 Key Electronics Developments, 1949-1995

Year	Product	Developer
1949	• Electronic tube	• Nanjing Electrical Lighting Factory
1951	• Wire recorder	• Shanghai Bell Sound Electrical Appliances Industrial Corp.
1952	• 15-watt short wave radio	• Tianjin Radio Factory
1956	• Total-nationalized electronic tube radio • Germanium alloy transistor	• Nanjing Radio Factory (sold internationally) • Inst. of Applied Physics/Chinese Academy.
1958	• B&W TV • Semiconductor radio • B&W TV broadcasting facilities • Electronic analogous computer	• Tianjin Radio Factory • Shanghai Hongying Radio Facilities Factory • Beijing Broadcasting Facilities Factory, others • Tianjin Electronic Instrument Factory
1959	• 500-gate automatic telephone switchboard • First successful silicon crystal bar	• Tianjin Zhongtian Motor Factory • Inst. of New Material Technology, Tianjin
1961	• DJS-1 digital computer: 30-bit wordlength, 30 times/sec, 1 K memory	
1964	• Silicon plane transistor • Epitaxial transistor	
1965	• Silicon digital integrated circuit	
1966	• Silicon bipolar ICs (in batch production)	
1967	• Huge transistor general digital computer	• Inst. of Computer Technology
1968	• PMOS and NMOS Ics	
1970	• China's first man-made satellite launched	
1971	• CMOS integrated circuit • Color TVs with PAL system	• 14th Radio Factory, Shanghai • Tianjin Radio Factory
1972	• General digital IC computer with a speed of 110 thousand times/sec	• Institute of Computer Technology, Shanghai
1974	• Doppler navigation radar and convective-layer-scattering communication machine • 128 K memory computer with 48-bit wordlength, speed of 1 million times/sec • 60 thousand voltage ion implantor	
1975	• 1 K DRAM • Hybrid analogous computer	
1977	• 800 K magnification electronic microscope	
1981	• Color video cassette recorder • Laser Chinese character editing. • Single—side density floppy disk	
1982	• 16 K DRAM	
1983	• Super high—speed computer.	• University of Defense Technology
1985	• 30 kW high frequency transmitter, key to positron-electron collisioner	
1986	• NMOS 64 K DRAM • 23 km optical fiber system in use	
1991	• 1M Chinese character ROM • Panda ICCAD system • 486 PC in batch production	
1992	• 900 MHz non-center addressing System • RISC workstation • Automatic customs declaration system • Weather monitor and forecast system • "Galaxy" supercomputer	
1995	• "Green Bird" large—scale software • Large—scale air transport control system • 1 mm VLSI in production	

Source: State Science and Technology Commission, 1997

4.1.8   The Ninth Five-Year Plan, 1996-2000

China's leadership recognizes the importance of electronics to the growth of the world economy, to individual national economies, and to its own national economic growth, scientific advancement, defensive capability, and social progress. China's electronics industry is therefore given special emphasis and support as a "pillar" industry, that is, a key industry that pushes forward the development of the entire economy.

4.1.8.1   Overall Development Goals

The overarching electronics goal of China's current Ninth FYP is to build its electronics industry into a major domestic economic sector capable of boosting China into a position of power in the global electronics and information technology marketplaces. To achieve this, the electronics industry is expected to sustain a 20 percent annual growth rate and achieve a gross production value of $85 billion (700 billion yuan) or 8 percent of the total national industrial output by the end of the century. Total earnings from sales of electronics products are expected to reach $72 billion (600 billion yuan), and exports are expected to reach $35 billion by the year 2000. This should place China among the top five electronics-producing nations of the world.

In striving to reach these ambitious goals, China is working to restructure its electronics industry from a traditional simple manufacturing system to a complex, multi-tiered modern system that features a combination of hardware, software, application, and information service capabilities, produced by a variety of types and sizes of enterprises, with large companies playing the leading role.[3]

The Ninth FYP emphasizes four electronics areas:

1.   semiconductors, especially integrated circuits;
2.   devices and components;
3.   computers and software;
4.   telecommunications and information technology.

Other areas of focus include process technologies, expanding national R&D capabilities, and building large enterprises comparable to the world's largest electronics firms. In addition, a series of "Golden Projects" aimed at modernizing China's national information infrastructure are also under way (see Section 4.3).

---

[3] Total production output of large enterprises, i.e., businesses with annual sales of more than $1.2 billion (10 billion yuan), is expected to comprise more than 60 percent of the industry's total output.

4.1.8.2 Size and Composition of the Industry

There were 3,417 enterprises in China employing about 1.7 million workers[4] in 1996 [*China Electronics Industry Yearbook 1997*, p. 11]. The bulk of China's electronics enterprises are located in Guangdong and Jiansu Provinces and in the Greater Shanghai City area. Table 4.3 shows the types of enterprises, their outputs and sales figures that comprised China's electronics industry in 1996. State-owned enterprises accounted for the largest share of output and sales of electronics followed by Chinese-foreign joint venture enterprises.

Table 4.4 presents China's major electronics product categories in terms of number of enterprises, total output values, and product sales in 1996. Radios and TVs remained the largest sector of China's electronics industry in 1996. Telecommunications equipment and electronics devices and components are the next two major sectors in terms of output value and total sales. Product categories with higher technical sophistication, such as electronic measuring instruments and semiconductor manufacturing equipment, accounted for much smaller shares of total output and sales.

Table 4.3
Types, Output, and Sales of China's Electronics Enterprises (1996)

Type of Enterprises	Number	Output Value ($ million)	Product Sales ($ million)
State-owned	1,378	9,281	8,942
Collective	1,096	3,229	3,082
Private	22	84	74
Joint-operation	48	308	297
Share-holding	108	1,628	1,598
Chinese-Foreign Joint Ventures	286	5,808	5,538
Chinese-Foreign Cooperation	17	39	38
Foreign-owned	19	2,741	2,709
Hong Kong, Macao, and Taiwan*	437	4,646	4,517
Other	6	14	13
Total	3,417	27,777	26,806

*Included joint ventures, cooperative agreements, and wholly-owned enterprises.
Exchange rate: US$1 = 8.30 yuan.
Source: *China Electronics Industry Yearbook 1997*, p. 12.

---

[4] Among the total workers employed by China's electronics industry in 1996, approximately 197,000 were engineers.

Table 4.4
China's Major Electronics Products (1996)

Products	Number of Enterprises	Output Value ($ million)	Product Sales ($ million)
Radar equipment	53	1,832	1,808
Telecommunications equipment	240	5,072	4,927
Radios & TVs	454	7,263	7,144
Computer	238	3,201	3,002
Electronics devices & components	1,325	4,628	4,388
Electronics measuring instruments	192	253	238
Electronics parts	257	512	504
Appliances	226	1,010	1,000
Electronics equipment	432	4,008	3,796
Semiconductor equipment manufacturing	248	384	365

Exchange rate: US$1 = 8.30 yuan.
Source: *China Electronics Industry Yearbook 1997*, pp. 18-21.

4.2   FOREIGN TRADE AND INVESTMENT IN CHINA'S ELECTRONICS INDUSTRY

Since the beginning of economic reform and opening to the outside world, China's electronics industry has achieved significant progress through trade and cooperation with foreign companies. This section first discusses Chinese-foreign cooperative ventures in China's electronics industry, followed by the structure and pattern of China's import and export of electronic products.

4.2.1   Cooperative Ventures in China's Electronics Industry

Between 1979 and 1982, approval and oversight of cooperative programs were strictly under the jurisdiction of the Central Government. Since 1983, some examining and approving rights have been given to the provincial and municipal governments. Before 1985, the guiding principle for Chinese-foreign cooperative ventures was unified planning and jointly dealing with the outside world in order to enhance technical quality, economic evaluation, and management. Since 1986, China has shifted from simply buying foreign equipment and know-how to allowing foreign companies to invest in China via contractual agreements, joint ventures, and foreign direct investments.

4.2.1.1 Cooperative Projects for the Period 1979 to 1982
Major Chinese-foreign cooperative projects in this period include:

Color TV sets and components
- three assembly lines for color TV sets: Tianjin Radio Factory with JVC; Shanghai No.1 TV Plant with Hitachi; and Beijing TV Plant with National (total cost, $20.35 million);
- ASICs for color TV sets: Jiannan Radio Apparatus Factory (now Huajing Electronics Group Corp. of China) with production techniques and equipment from Toshiba (cost, $65 million; production capacity, 26.48 million chips a year);
- color kinescopes: Shangxi Color Kinescope Factory (now Color Rainbow Group Corp. of China) — complete production techniques and equipment for color kinescopes, glass shells, and masking technologies transferred from National and other Japanese firms (cost, $156 million; production capacity, 0.96 million per year);
- assembly line for color TV scanning transformers (cost, $4.8 million; production capacity, 1 million per year)
- assembly line for high-frequency modulators for color TV sets (cost, $3.7 million; production capacity, 2 million per year);
- eighteen production lines purchased from abroad for electronic devices and electric components for use in production of color TV sets, including five lines for discrete semiconductor devices, thick and thin circuits, and speakers and their components (total cost ~$50 million);
- printed circuit board (PCB) production line purchased from abroad: Shanghai No.20 Radio Factory (cost, $3.8 million; production capacity, 0.15 million square meters a year to meet production requirements of a million color TV sets a year).

Receivers and Radios
- six audio cassette recorder and radio assembly lines (capacity, >0.8 million sets a year;
- three audio cassette recorder core production lines (capacity, 1.8 million cores a year);
- one recorder magnet tip production line (capacity, 4.8 million tips a year)

Black & White TV sets
- five B&W TV kinescope production lines (capacity, 5.52 million B&W kinescopes per year to meet production requirements of 5.5 million B&W TV sets a year).

4.2.1.2 Cooperative Projects for the Period 1983 to 1986
It was decided in April 1983, during the national conference on technical improvement of machinery and electronics industries jointly held by the State

Planning Commission and the State Economics Commission, that provincial and special municipality governments as well as relevant ministries, should be given more rights to introduce foreign projects. They were given the right to approve projects up to $5 million. The Shanghai, Tianjin, and Dalian municipal governments, were given the rights to approve up to $30 million, $20 million, and $10 million, respectively. By the end of 1985, more than a thousand contracts costing more than $10 million had been signed with foreign companies. The major projects approved in this period included:

- 113 color TV assembly lines with a manufacturing capacity of 15 million sets per year;
- a number of combination production lines of glass shell, devices, and materials for production of color TV sets;
- $218 million spent by thirty-three enterprises to purchase manufacturing technologies and equipment for production of audio/video recorders, communications and navigation instruments, fiber optics, personal computers, IC chips, and discrete semiconductor devices.

4.2.1.3 Cooperative Projects since 1986

Chinese-foreign cooperative projects or joint ventures in the electronics industry during this period included.

IC Chips
- Shanghai Belling Microelectronics Manufacturing. Corporation was jointly established by Shanghai No.14 Radio Factory and Belgium Bell Company.
- Advanced Semiconductor Manufacturing Corp. of Shanghai was jointly established by Shanghai No.5 Components Factory, Shanghai No.7 Radio Factory, and Philips Co. of the Netherlands.
- ShouGang NEC Company was jointly established by Capital Iron-Steel Company of China and NEC Co. of Japan.
- Hua Hong NEC Company was jointly established by Shanghai Hua Hong Microelectronics Company and NEC Co. of Japan in October 1997.

Communications
- Shanghai Bell's major products include program-controlled telephone switchers with a capacity of 0.3 million channels per year. An investment of $150 million by Pudong Shanghai Bell has increased its capacity to produce 2.1 million channels of new switchers annually.
- Beijing Wired Electric Equipment Factory of China, a joint venture established with Germany's Siemens, had an original capacity of 0.3 million channels of program-controlled telephone switchers per year.
- Tianjin Tianzhi Communication Company, a joint-venture established with Japan's Toshiba, had an original capacity of 0.3 million channels of

program-controlled telephone switchers. As of 1997, the capacity has increased over 1 million per year.

- Two program-controlled switcher factories, one in Qingdao and one in Chendu. Both were joint ventures with AT&T, started in 1993. (Capacity was three million each).
- Two-million-channel factory was established by Shunde in Guangdong province with Northern Communication Company of Canada.
- $50 million was spent by ten Chinese factories to transfer technologies from foreign companies; each of the factories has an annual production capacity of approximately 100,000 units:
  - Hunan Changde Wired Apparatus Factory, with Philips (the Netherlands);
  - Suchou Wired Apparatus Factory, with Philips;
  - Huayun Program-Controlling Dept. of Zhenghua Co., with Philips;
  - Beijing Wired Apparatus Factory, with Ericsson (Sweden);
  - Tianjin Tianzhi Comm. Company, with TAI (USA);
  - Shanghai Xinguang Communications and a Beijing factory, with Siemens (Germany);
  - Guangzhou Wired Apparatus Factory, with Harris (USA);
  - Tongguang Electronic Communications and Jiangxi Wired Apparatus Factory, with NORTEL (Canada);
  - Shanghai Telephone Apparatus Factory, with Plessy (U.K.);
  - "Feb.7" Comm. Factory of the Ministry of Railways;

Fiber-optic communications equipment

- Zhongqing factory (under the jurisdiction of Ministry of Posts and Telecommunications) and Sichuan Peijiang Wired Electric Instrument Factory transferred technologies and production lines from Yidatire Company (Italy) and NEC Company (Japan).
- Tianjin Electro-Optical Equipment Co. introduced a production line of one to four optical-electric group terminals from Seimens (Germany).
- The Communication Equipment Company was established as a joint venture by the Shanghai government and AT&T (USA) to produce four optical-electric group terminals.
- China's Ministry of Posts and Telecommunication and Fujitsu (Japan) jointly established a development and production base in Wuhan to manufacture optical fiber and cable. Its production capacity was 100,000 kilometers of optical fiber per year.

Mobile communications

- Nanjing Radio Factory (China) and Ericsson Company (Sweden) jointly produced wireless switchers, base stations, and mobile stations in China with an investment of more than $40 million.

- Motorola (USA) established a wholly-owned plant in Tianjin to produce assigned circuits and mobile communication apparatus.

Satellite communications

- A number of foreign companies, such as AT&T, Hughes, NEC, NCR, and Atlanta, have participated in projects to establish a professional communications network in China for Civil Aviation, Customs, the Ministry of Communications, the People's Bank of China, the Xinhua News Agency, and some other government departments.

Videocorders

- Dalian (Group) Corp. (China), which was founded in 1991, purchased production techniques from National (Japan) for $188 million (1 billion yuan) to manufacture three million component sets of video recorder cores and magnetic drums per year.
- Tianjin Samsung. was founded in 1992 as a joint venture of a Tianjin factory with Samsung Company of Korea. As of 1997, it has a production capacity of one million video recorders per year.

4.2.1.4   *China's Utilization of Foreign Capital in the Electronics Industry*

Most foreign investments in China's electronics industry are in the form of joint ventures. The number of foreign contracts and the volume of foreign capital invested in China's electronics industries have increased over the years (Table 4.5). Table 4.6 lists the sums of utilized foreign capital and their percentages of total investment in selected municipalities and provinces. Table 4.6 shows that during the decade between 1985 and 1995, foreign investment in China's electronics industry utilized in Tianjin and Shanghai increased by huge amounts. In addition, foreign investment in the electronics industry as a percentage of total national investment has increased annually. In 1985, restricted by government policies, the ratio of foreign investment in the electronics industry to total investment could not exceed 50 percent. However, the restriction relaxed in the 1990s. As a result, the ratio reached 67.1 percent in Beijing and 55.3 percent in Tianjin. Table 4.7 summarizes foreign capital invested in China's electronics industry in 1995. In terms of utilized foreign capital, state-owned enterprises remained the top category in foreign cooperation, followed by direct investment in China by foreign multinational corporations.

Table 4.5 Foreign Investment in China's Electronics Industries

Year	Joint Ventures	Foreign Capital Contracted	Foreign Capital Utilized
1986	150	0.2	
1988	536	0.6	
1991	2,846	2.5	
1993	5,000	4.2	
1994	8,000	7.0	4.0*
1995	10,246	9.5	7.0*

Unit: US$ billion   *Accumulated number
Source: State Science and Technology Commission, 1997

**Table 4.6 Utilized Foreign Capital and Percentage of Total Investment in
Selected Provinces and Special Municipalities in China**

Location	1985		1990		1993		1995	
	Utilized Foreign Capital*	% of Total Investment	Utilized Foreign Capital*	% of Total Investment	Utilized Foreign Capital*	% of Total Investment	Utilized Foreign Capital*	% of Total Investment
Beijing				39.8		55.1		67.1
Tianjin	1.2	31.3	37.4	48.1	144.7	49.2	265.3	55.6
Liaoning	2.9	32.8	12.2	19.6	37.6	38.5	83.8	43.1
Shanghai	8.2	45.9	82.8	31.8	493.4	42.8	1,699.3	55.3
Jiangsu	0.9	1.0	41.0	28.0	64.8	45.0	50.5	40.0
Shangxi	50.0	50.0	4.2	50.0	18.4	42.0	48.7	43.0

Source: State Science and Technology Commission, 1997.   *$ millions

Table 4.7 Foreign Capital Invested in China's Electronics Industry, 1995 ($ millions)

Type of Enterprises	Contracted Foreign Investment	Utilized Foreign Investment
State-Owned	436.2	345.3
Collective	49.7	31.7
Intensive	11.7	41.7
Stock Companies	228.0	63.8
Joint Ventures	244.6	184.9
Joint Cooperative	4.6	5.5
Foreign-Owned	239.5	237.0
Joint Venture with Hong Kong, Macao, and Taiwan	44.3	52.8
Joint Cooperation with Hong Kong, Macao, and Taiwan	5.2	30.1
Owners from Hong Kong, Macao, and Taiwan	49.8	45.2
Total	1,313.5	1,037.9

Source: State Science and Technology Commission, 1997

As of 1995, foreign investment in China's electronics industry could be summarized as follows:

- Over half of all foreign investment (53.5 percent or $555.4 million) was utilized by equity joint ventures, cooperative joint ventures, and foreign-owned enterprises.
- Among domestic enterprises, state-owned enterprises still dominate utilization of foreign investment.

The investment of $237 million from foreign-owned enterprises comes from only six firms, which shows that big foreign companies dominate foreign investment in China's electronics industry.

- The actual percentage of utilized foreign investment in equity joint ventures, cooperative joint ventures, and foreign-owned enterprises (92.8%) is considerably higher than that in domestic enterprises as a whole (66.5%).

4.2.2 Imports and Exports of Electronics Products

China's electronics products made an international market debut in 1956. Initially, exported electronics products were limited to radio components (resistors, capacitors, coils, earphones, speakers) and some lighting instruments. The variety was limited and the quantity was low, bringing only several hundred thousand U.S. dollars a year. Hong Kong and Macao were the only markets. Around 1963, radios and low-end assembly products were added, and the market was expanded to Singapore and Malaysia. It the same period, China exported components and materials to the Soviet Union, Eastern Europe, Mongolia, North Korea, Vietnam, and Cuba on a barter basis. The volume of foreign trade during the Guangzhou Fair in Spring 1964 was only $102 million. In 1965, the exported volume of China's electronics products was about $2 million, which was 0.34 percent of the domestic electronics industry's output value and 0.1 percent of China's total export volume. During the Cultural Revolution there was no growth in the electronic products. After 1972, export volume began to grow, and in 1976, was $10 million, the highest in history, but it then fell abruptly from 1977 to 1979.

In the 1980s, economic reform and open-door policy brought great vitality to China's electronics trade. Approved by the State Council, the China National Electronics Import and Export Corporation (CEIEC) was established in February 1980. As a result, a series of policies encouraging exports was formulated, and many export bases were set up. The structures of export products have experienced great change. In the 1980s, the main export products were only basic components and devices. In recent years, complete sets of equipment or products have constituted 80 percent of total export value. Many high-tech products, such as computers, satellite antennas, and program-controlled telephone and switching systems, have been exported as well.   Export volume was $18.51 million in 1981, which grew to $100 million in 1986, $200 million in 1987, $1 billion in 1992, and $1.3 billion in 1994. China attained a favorable balance of trade in the electronics field in 1995 [CEIEC, 1997].

4.2.2.1   Patterns of Import and Export Electronics Products

An export orientation is a salient feature of China's electronics industry, which leads the nation in manufacturing products trade. Table 4.8 displays China's total value of electronics trade for 1994-1996.   Total electronics imports to China were $18 billion in1996, 13 percent of the nation's total imports ($138.8 billion). Imports of electronics products increased 11.3 percent over 1995. The 1996 value of China's electronics exports was $21.5 billion, 14.2 percent of the nation's total exports ($151.1 billion). The total exports of electronics products increased 30 percent over 1995.

Table 4.8 shows China's imports and exports classified by category of products. The import value of basic components and devices exceeded the import value of audio/video and communications products in 1996, while audio/video and communications products remained the top category of electronics exports in 1996. Table 4.9 presents China's electronics trade by type of enterprise. State-owned enterprises accounted for the largest proportion of China's electronics import and export activities, followed by Chinese-foreign joint-venture enterprises and foreign-owned businesses. Table 4.10 displays China's electronics trade by method of trade. Material processing and assembly was the most important trading method in China as of 1996. It indicates that the majority of China's electronics industry still remains simple and labor-intensive.

4.2.2.2   China National Electronics Import and Export Corporation

The China National Electronics Import and Export Corporation (CEIEC) is the main import/export channel for China's electronics products. Established in 1980, CEIEC is a foreign trade company integrating trade with manufacturing and electronics technology. Its total import/export volume ranked seventh and its export volume fifth on the list of large-scale foreign trade enterprises of China from 1993 to 1995. The average growth rate

Table 4.8

China's Electronics Imports and Exports by Product Category

Product	1994		1995		1996	
Category	Import	Export	Import	Export	Import	Export
Basic Components/ Devices	4,456	2,379	6,540	4,280	9,345	6,414
Audio/Video and Communication Products	6,861	6,931	6,963	7,555	8,543	8,938
Computers	2,162	3,052	2,653	4,697	3,087	6,146
Total	13,478	12,362	16,156	16,532	17,976	21,498

Unit: US$ million

Source: *China Electronics Industry Yearbook 1997*, pp. 234-235.

Table 4.9

China's Electronics Imports and Exports by Type of Enterprise

Type of Enterprise	1995		1996	
	Import	Export	Import	Export
Chinese-foreign cooperations	873	1,255	1,078	1,404
Chinese-foreign joint ventures	4,348	4,877	5,791	7,088
Foreign-owned enterprises	3,437	4,359	4,973	6,899
State-owned enterprises	7,159	5,861	5,809	5,814

Unit: US$ million

Source: *China Electronics Industry Yearbook 1997*, pp. 234-235.

Table 4.10

China's Electronics Imports and Exports by Method of Trade

Method of Trade	1995		1996	
	Import	Export	Import	Export
General Merchandise Trade	5,368	1,917	4,130	1,543
Material Processing & Assembly	2,432	3,802	2,825	4,635
Imported Material Processing	7,231	10,491	8,970	14,904
Imported Equipment	225	N/A	143	N/A
Equipment Imported by Foreign-owned Enterprises	480	N/A	1,085	N/A
Duty-Free Merchandise	283	N/A	264	N/A
Others	239	4	559	416
Total	16,156	16,532	17,976	21,498

Unit: US$ million

Source: *China Electronics Industry Yearbook 1997*, p. 235.

of its import/export volume was 20.4 percent in the period from 1990 to 1995. In 1995, its export volume ($1.4 billion) was 6.6 percent of all Chinese exported electronics products, and its import volume ($1.1 billion) was 8.4 percent of all electronics products imported to China. As of 1997, the number of CEIEC's subsidiaries has increased to forty-six, which are scattered all over the country in provinces, autonomous regions, and cities like Beijing (CEIEC Beijing), Shanghai (CEIEC East China), Shenzhen (CEIEC Shenzhen), and Guangzhou.

In the mid-1990s, CEIEC's business scope expanded beyond electronics to other sectors, such as machinery, light industrial products, and chemicals. Facing the intense competition in the global marketplace, CEIEC has set up thirty various trade companies and joint ventures in countries including the United States, Germany, Thailand, Brazil, Japan, South Africa, and Russia. All of them have acted to enable CEIEC and domestic enterprises to expand their foreign economic cooperation and trade.

4.2.2.3 Prominent Features of China's Trade in Electronics Products

China's international trade in electronics products has the following prominent features:

Growth of exports and decline of imports in electronics product trade is tied to the development targets of the industry. Before 1988, the volume of China's electronics imports was much greater than its exports because the industry was not competitive in the global marketplace. After the electronics industry was designated to be one of China's pillar industries, resources and policies have been focused on the developing of industry. The quality, variety, and sophistication of China's electronics products has improved, and the growth in export volume has increased. In 1989, the ratio of import volume was 2:1. During the Eighth Five-Year Plan (1991 – 1995), the output value of the electronics industry in China was growing an average of 28.7 percent annually, which was 2.4 times the average growth of the gross national product (GNP) in the same period. Figure 4.1 shows China's import and export values of electronic products since 1986.

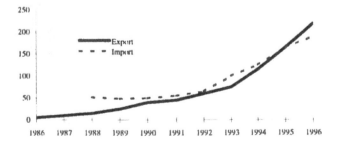

Figure 4.1 Import and Export Value of China's Electronics Industry ($100 million).

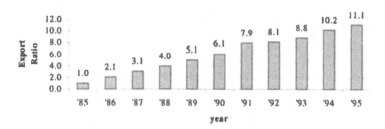

Figure 4.2 Ratio of Export of Electronic Products to Total Trade Volume.

1. The ratio of export electronics products to total national trade volume is growing annually. The export ratio of electronics products to total national foreign trade has been growing annually since 1985. Figure 4.2 shows that this ratio increased eleven times in the decade 1985-1995.

2. The growth of trade volumes in electronics products is faster than that of all products as a whole. Table 4.11 shows that the growth speed of import and export volumes of electronic products was 6 percent and 17 percent higher than that of the total national foreign trade, respectively, in the period of 1990-1995. The average growth rates of both import and export volumes in electronics products were also significantly higher than those of the national average growth rates in foreign trade.

3. The structure of exported electronics products has been improving. According to Chinese Customs figures and as shown in Figure 4.3, the share of exported consumer electronics products, mostly video and audio products, has decreased, and the share of investment electronics products, mostly computers and basic components and devices, has increased annually since 1994.

Table 4.11
Comparison of Growth Rates of Total National Foreign Trade and Electronic Products
Foreign Trade, 1990 to 1995

	Average Growth, Total National Foreign Trade	Average Growth, Electronic Products Foreign Trade
Export and Import Volume	16.8%	27.1%
Import Volume	15.1%	21.4%
Export Volume	19.2%	36.1%

Source: State Science and Technology Commission, 1997.

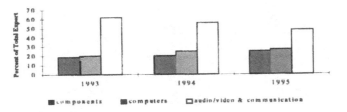

Figure 4.3 The Proportion of Various Electronics Export Products to Total Electronics Trade.

4. <u>The percentage of exported electronics products made in joint ventures or foreign-funded enterprises is increasing annually</u>. Encouraged by the new national policies, foreign capital introduced by contract has topped $7 billion, and more than eight thousand enterprises have utilized foreign investment. The export volume of foreign-funded enterprises[5] is growing annually and has accounted for more than two-thirds of total export volume (see Table 4.9).

5. <u>China's exported electronics products mainly consisted of processing products made with imported materials or materials provided by foreign companies.</u> About 80 percent of China's electronics exports were merely processed and assembled in China (see Table 4.10). In addition, the volume of re-exported products after processing is increasing. This indicates that the general level of sophistication of China's electronics industry is still fairly low, and most exported electronics products are still labor-intensive.

4.3 MAJOR NATIONAL ELECTRONICS PROJECTS

National planning of major electronics projects for the ninth Five-Year Plan (FYP) includes several "Golden Projects" that aim to modernize the country's information technology infrastructure. The Golden series projects include the nationwide public economic information processing network (Golden Bridge Project), the electronic monetary and modern payment system (Golden Card Project), the foreign trade information sources network (Golden Customs Projects), the electronic taxation system (Golden Taxation Project), the industrial production and circulation information network (Golden Enterprises Project), education and research network (Golden Intellectual Project), agricultural management and service network (Golden Agriculture Project), and national economic micro-policymaking support system (Golden Policy Project). In addition, long-range national planning in the electronics industry also includes a semiconductor project and two major electronic system projects. This section introduces four major Golden series projects, the 909 semiconductor manufacturing projects, the air traffic

---

[5] Including contractual agreements in Chinese-foreign enterprises, Chinese-foreign joint-ventures, and foreign-owned enterprises.

control system project and the Three Gorges Dam electronics systems project.

4.3.1 China's "Golden Projects" for Information Management

As part of its Ninth FYP, the Chinese government has embarked on a series of "Golden Projects" as backbone projects in support of national economic and social development. Four major Golden Projects are intended to advance and popularize the applications of electronics technology in the areas of macroeconomic control, finance, international trade, and tax collection.

4.3.1.1 Golden Bridge Project

On March 12, 1993, the National Public Economic Information and Telecommunication Network – Golden Bridge Project was mapped out in a conference presided over by Vice Premier Zhu Rongji. The Golden Bridge Project is China's version of the information superhighway. The purposes of this project are to serve state macroeconomic control and strategic decision-making, to facilitate the sharing of national economic and social information, and to build and promote the development of a modern electronics information industry.

As part of China's information infrastructure, the Golden Bridge Net has been interconnected with the Ministry of Posts and Telecommunications and other departments, and has formed a medium-speed information and telecommunications network covering all of China. It will offer an integrated telecommunications system through satellite and optical fiber networks for a variety of information service systems, such as banking, customs, foreign trade, domestic trade, traveling, weather, transportation, agriculture, irrigation works, forestry, education, research, and development.

JiTong Communications Co., Ltd., a joint stock company owned by a consortium of twenty-six shareholders affiliated with MEI, was responsible for initiating the Golden Bridge Project with a satellite truck network in March 1993. At the end of 1994, the basic satellite system of the Golden Bridge Network had been completed. By 1996, the Golden Bridge network had gone into operation in twenty-four provinces and cities and was interconnected with CERNET, the State Information Center, the Information Center of State Economic and Trade Commission, and the National Electronic Press Service Center. At the same time, China developed an EDI/E-mail exchange platform and a high-speed broadband transmission system.

Major goals that remain to be accomplished include constructing a terrestrial network that will broaden bandwidth and improve availability, promoting the Internet service and value-added service nationwide, and improving electronic commerce applications in major cities.

4.3.1.2 Golden Card Project

In June 1, 1993, while visiting Shahe General Satellite Clearing Center of the People's Bank, Jiang Zemin, the general secretary of the Communist Party of China, announced that credit cards should be used by the people in order to reduce the amount of cash in circulation and enhance national macroeconomic control. This started China's "Golden Card" project, which emphasizes the development of magnetic card technology and its applications. The Golden Card project will serve government agencies, banks, the postal service, telecommunications, domestic trade, and tourism by making full use of existing communication network resources such as the Golden Bridge network. The project will popularize credit cards and cash cards that will enable people to make electronic cash deposits, withdrawals, and payments. At the same time, new electronic payment methods will be used in business and trade systems and the tourist industry. The goal is to use telecommunications networks to replace cash transactions to increase people's convenience, comfort, and access to information. The Golden Card Project has begun with banking cards and some other card-based application systems. By the year 2005 the plan will deliver three hundred million credit/cash cards to three hundred million people in four hundred urban areas.

In October 1993, the State Office for the Golden Card Project, comprised of the People's Bank of China, the Ministry of the Electronics Industry, the Ministry of Posts and Telecommunications, the Domestic Trade Department, and the National Tourist Bureau, was founded within the MEI.

China is aggressively developing the Golden Card Project. Various memory cards, IC cards, and other active or passive radio cards are widely used in fields such as transportation, telecommunications, and personal identification. Credit cards were first issued in China by five commercial banks in 1986. Four million cards had already been issued when the Golden Card Project started. The average number of cards issued during the first three years of the Golden Card Project was seven million per year. The card growth rate was 183 percent. As of the end of 1996, there were 200,000 business credit card accounts, 10,000 ATMs, and 70,000 points-of-sale (POSs) in China.

Among the twelve provinces and cities experimenting with the Golden Card Project, eleven have succeeded in using ATMs and POSs by interconnecting themselves to the Golden Bridge Network (CHINAGBN). For example, in Shanghai as of mid-1997, there were 1,019 ATMs, 95 percent of which had been interconnected to the Net; 4,400 POSs; 2,000 business users; and 200 net-interconnected units. There were 3.5 million bankcards issued by banks, among which 300 thousand were IC cards.

In June 1996, discussions were under way to interconnect the Golden Card Project with the international VISA net so that people could withdraw yuan from Shanghai ATMs with VISA cards. The ATM net in the Jiangsu

Province opened in December 1996. About 100 ATMs in twelve banks from the city of Nanjing, Wuxi, and Suzhou have been interconnected.

Non-banking card applications now include organization identification cards, taxi IC cards, auto transportation charging systems, the Shanghai tax reporting system, gasoline IC cards in the city of Qingdao, a housing funds system in Tianjin, and an electronic wallet electronic bankbook system in Guangdong. It is estimated that the annual demand for IC cards will reach four hundred million in China by the year 2000. As of 1997, there were thirty credit card assembly and packaging lines in China, with a total production capability of two hundred million cards per year. There were also five processing lines for IC cards that produced six hundred million chips annually.

4.3.1.3   Golden Customs Project

In 1993, an information service net, the Golden Customs Project, was established to connect foreign trade companies with banks and China's customs and tax offices. This project aims to create paperless trading by automating customs checks and eliminating cash transactions for international trade. Its main tasks include

- establishing a foreign trade information services network interconnected with government departments such as customs, commodity inspection, tax revenue, the Ministry of Foreign Economy and Trade, the National Statistics Bureau, Foreign Currency Administration, and banks and foreign transportation and import/export enterprises through the Golden Bridge Net
- establishing an information application system for export tax-return management, foreign currency exchange and clearing, and maintenance of import and export statistics;
- establishing an information exchange service center with branches;
- achieving standardization for information exchange, bills, and certificates in order to enhance and improve foreign trade management;
- supporting the realization of paperless trade by establishing experimental units for EDI application, nationwide E-mail service, and an electronic post office.

As of 1998, the Information Exchange Service Center was completed, and an e-mail platform is in operation. The Imports and Exports Permission Auditing system and the exports tax return system are partly operational. The checking system for imports and exports of the Foreign Currency Administration has achieved interconnection with the Customs Department and the General Bureau of Taxation. The Imports and Exports Statistics System has been interconnected with the Ministry of Foreign Economy and Trade.

4.3.1.4 Golden Taxation Project

In 1994, Vice Premier Zhu Rongji decided that China should establish a computer network system for tax collection, which was later named the Golden Taxation Project. This is a nationwide information management project to enhance tax collection and management and prevent losses of tax revenue due to tax evasion. The basic infrastructure was completed in August 1994 in more than fifty cities and counties, and the application software runs in the network of the General Clearing Center of the Peoples Bank of China.

As part of the project, the General Company of the Aviation Industry developed a value-added, tax invoice counterfeit-proof system. An electronic tax declaration system, developed by Great Wall Computer Corporation, has been tested in Shanghai since August 1995. And a tax-count cashier machine system has also been tested and used in the field since 1996.

4.3.2 909 Project

The 909 Project (see Chapter 5) is the most expensive semiconductor engineering project in China's history. The objective of the project is to provide a major boost to domestic manufacturing capability by establishing a domestic semiconductor industry that can eventually meet the rapidly increasing domestic demand for high-technology devices. The project began in the Ninth FYP in 1996. This project includes an 8-inch silicon wafer processing line, manufacturing plants, and an IC foundry facility. The General Company of Ferrous Metals of China is in charge of manufacturing the silicon wafers. The HuaDa IC Designing Center is responsible for designing the plant. Many other IC design units will take part in designing IC products and services.

The Hua Hong Microelectronics Company, Ltd., was founded at the end of 1995 when China's State Council decided to establish an 8-inch wafer processing line in Pudong, Shanghai, and selected NEC as the foreign partner. The goal is to build a $1 billion IC production line in the Shanghai Pudong Jingqiao Development Zone. The plant will be China's largest IC production facility, capable of manufacturing 20,000 8-inch silicon wafers per month. China currently has the capability to produce chips with a width of 1-2 µm, and this new facility will bring China's process capability to 0.35 to 0.5 µm.

The registered capital of Shanghai Hua Hong NEC Electronics Company, Ltd., is $700 million, of which $500 million (71.4 percent) is from Shanghai Huahong Microelectronics Company, Ltd., $130 million (18.6 percent) is from NEC Japan, and the rest is from NEC China. The general manager reports to a board consisting of ten members, seven of whom are from China and three of whom are from Japan. Zhang Wenyi, vice minister of MEI, is the director of the board, and the deputy director is Zou Shichang from China. The general manager is from Japan. The joint venture will achieve full-scale volume production in mid-1999.

In March 1998, NEC signed another joint-venture agreement with China to design and sell semiconductors. Beijing Hua Hong NEC IC Design Co. Ltd., will be established in Beijing in June 1998 and begin operations in January 1999. The new joint-venture company strengthens NEC's commitment to participation in China's 909 project and broadens China's semiconductor industry by adding state-of-the-art design capabilities. Overall investment in the joint-venture company will amount to $30 million, with capitalization of $20 million, 41 percent contributed by NEC, 10 percent from NEC (China) Co., Ltd., and 9 percent from Shougang NEC for a total of 60 percent from the NEC Group, and the remaining 40 percent from Chinese partner Beijing Hua Hong IC Design Co. Ltd. [*NEC Press Release* March 1998].

4.3.3 Electronic Air Traffic Control Project

During the Ninth Five-Year Plan, China plans to invest 9 billion yuan ($1.1 billion) to build a state-of-the-art electronic air traffic control system that will be functional by the year 2000 in eastern China. The system will combine ground radar control with an air navigation system to ensure air safety.

China is installing over three hundred field satellite stations, three hundred air-oriented transceivers, fifty meteorological radar apparatuses, and 60 sets of satellite cloud picture receivers. At the same time it is working to establish a meteorological data library. The system will include both land-based systems and satellite-based. Support technologies under development include solid-state linear radar, single-pulse quadratic radar, modern telecommunications net technology, computer net technology, and multi-radar comprehensive information processing technology.

4.3.4 Three Gorges Electronic System Project

The Three Gorges Dam on China's Yangtze River is a huge national engineering project on which actual planning began in the 1980s and construction began in 1994; it is scheduled to take twenty years to complete. If built according to plan, it will be the largest hydroelectric dam in the world, almost a mile wide and 575 feet high, above the world's third longest river. The project includes a massive power generation facility, locks for diverting shipping, and a reservoir that will stretch over 350 miles upstream and force the displacement of 1.2 million people. The plans have been controversial both internally and abroad because of the necessary displacement of people, environmental destruction, and engineering problems in the steep river gorge where the dam is to be built.

The Three Gorges power station is designed with twenty-six hydraulic generators with a capacity of 700,000 kW each and a total capacity of 1.82 million kW. An essential component of the plan is a comprehensive state-of-the-art electronic systems management project, with five components:

1. *The Three Gorges Project Information Management System* is a unified information services system that can assist in the whole process of design and management of the Three Gorges Dam, water control, and power generation project.

2. *The Three Gorges Telecommunication System* provides communications support for the power station, electrical power system, flood control, and shipping management.

3. *The Three Gorges Power Station and Power Transmission Automated Monitoring System* is the heart of the Three Gorges key water control project, which will monitor and manage the operation of the power station, substations and transmission system; conduct data collection; process, monitor, and operate all equipment; and deal with accidents and breakdowns

4. *The Navigation Control System* manages shipping activities and ensures navigation safety in the Three Gorges area, as well as publishing shipping information, weather and hydrological reports, and administrative instructions.

5. *The Weather and Hydrological Information System* includes meteorological radar, weather satellite receiving systems, hydrological information collection and process systems, and a weather information network.

Uncertainty about the dam's technical, environmental, social, and financial feasibility has caused many traditional dam financiers, including the World Bank, the U.S. Export-Import Bank, and the Tennessee Valley Authority, to give the project a wide berth. China has tried to sidestep foreign concerns by funneling foreign financing for the project through the State Development Bank of China, a so-called "policy bank" created in 1994 specifically to fund the Three Gorges and other large politically motivated projects.

The total official estimated cost of this project is $29 billion. Half of it will come from surcharges on electricity bills, and 12.5 percent is a loan from the State Development Bank of China. The remainder is to be raised from bond issues and loans from other domestic and foreign banks.

Chapter 5

CHINA'S SEMICONDUCTOR INDUSTRY

Domestic demand for semiconductors, especially integrated circuits (ICs), is growing exponentially in China, stimulated by the development of information technology and advanced telecommunication infrastructures throughout the country, and by growing demand for common consumer products. Although it is difficult to gauge the exact number of ICs being imported, various estimates of China's semiconductor market (e.g., by Dataquest, the Japan Electronics Industries Association, and SEMI) indicate that the market size was about $5 to 7 billion in 1996 and $8 billion in 1997, and will be about $15 to 17 billion by the year 2000 [Lammers 1997, Koo 1997].

China is competing with its Asian neighbors as well as with technologically advanced Western companies. In terms of 1995 global market share, U.S. and Japanese companies together accounted for 80 percent of the world semiconductor market, while four Asian IC manufacturers, Korea, Taiwan, China, and Singapore, accounted for 12 percent of the world market [Simon 1996]. Competition among Asian rivals has intensified since the early 1990s. South Korea's top three semiconductor chip manufacturers, Hyundai Electronics Industry, LG Semicon, and Samsung Electronics Company, each boasted profits over $1.3 billion in 1995 [Simon 1996]. In 1997, Samsung was ranked third among chipmakers worldwide and first among DRAM (dynamic random access memory) producers. In addition, Taiwan Semiconductor Manufacturing Corp. (TSMC) is the world's largest IC foundry company.

In contrast to the highly sophisticated production facilities of Korea and Taiwan, China's semiconductor industry still consists of relatively small-scale manufacturers with low productivity and low-level process technology. Poor infrastructure and immature peripherals industries contribute to inefficiency in the industry. Despite these limitations, China's semiconductor industry is expanding rapidly. It is expected that, by the year 2000, China alone will account for 15 percent of the total Asian demand for semiconductors, with projected annual sales of over $2 billion [Simon 1996].

5.1   INTEGRATED CIRCUITS

China's IC industry was virtually nonexistent before 1980. In 1996, China produced less than one percent of the world's ICs. But the strength of the country's growing electronics sector —already a major exporter —assures a ready market for any suitable IC that Chinese wafer fabrication plants can produce. Help for this budding semiconductor industry is coming from global companies such as Motorola, NEC, Philips, Siemens, and Toshiba. These companies are transferring technology, building wafer fabs, and forming joint ventures with Chinese partners.

5.1.1   Government Goals for China's IC Industry

During China's Eighth Five-Year Plan (FYP) period (1990-1995), the electronics industry experienced rapid increases in productive capacity, technological capability, electronics output, and international trade volume. By the end of the Eighth FYP, the total output of China's electronics industry reached $30 billion [Schumann 1997]. In its Ninth FYP (1996-2000), the Chinese government is anxious to further increase the production capability of its domestic IC industry. China's government expressed its goals for IC production in broad terms [Schumann 1997]:

- reach large-scale production levels for 6 inch and 0.8 µm process technology;
- enter industrial production for 0.5 µm and 8 inch wafer technology;
- increase IC design capability to meet market demands;
- pursue R&D in 0.3 to 0.4 µm and advanced packaging technology;
- develop 8 inch, single-crystal wafer technology and begin domestic production.

In order to achieve its goals, China must rely on foreign technological know-how, while at the same time taking steps to protect its large market from foreign domination. Therefore, many restrictions are imposed on Chinese-foreign joint ventures as well as on wholly-owned foreign enterprises to guarantee a certain level of technology transfer to China and to promise a significant portion of output for export.

5.1.2   Project 909

As a major part of China's Ninth FYP, to encourage domestic IC production capability and to reduce reliance on semiconductor imports, in December 1995 China launched its largest ever IC development project in the Pudong New Area of Shanghai. With an investment of more than $1.2 billion, this project is the largest project ever undertaken in China's electronics industry. The Pudong microelectronics center enterprise is just one piece of a larger project known as Project 909, sponsored by China's Ministry of the Electronics Industry (MEI). Shanghai was chosen as the preferred site for the project because it has become the center for microelectronics production in China. In 1995, Shanghai plants accounted for 21 percent of total Chinese production of semiconductors [Simon 1996]. Project 909 will start with the development of an 8 inch, 0.5 μm single-crystal wafer manufacturing facility, followed by several others and a number of IC design and development facilities [Schumann 1997]. Volume production is scheduled for the first half of 1999. The ultimate good of this project is to bring 0.35 micron, very large-scale integrated circuit (VLSI) technology to bear on telecommunication and computer-use ICs [Lammers 1997].

5.1.3   IC Production and Market Size

China's total output in IC products reached close to 760 million units in 1996, which accounted for less than 0.5 percent of total world production [Weng 1996]. In terms of technological sophistication, China first applied 5 μm process technology, which was the technological level in the U.S. and Japan during the early 1970s, to manufacture IC products in 1986. In 1994, China was able to improve its IC mass production process capability to 3 μm and applied to two MOS LSI production facilities in Hua Jing Electronics Group [Weng 1996].

In 1995, the Chinese government approved a total of 2,150 million yuan ($257 million) to build several VLSI plants. Of this, 1.4 billion yuan ($168 million) was spent for Hua Jing to construct a single 0.8 ~ 1.0 μm VLSI product line. The monthly output capacity is expected to be ten thousand pieces of 6 inch wafer and more than fifty varieties of IC products [Weng 1996]. In 1997, of some three thousand different kinds of IC products, most were small-scale ICs using 0.8 μm processing technology, which satisfied about 10 percent of total domestic demands. The technology in 1997 was 0.8 μm on 4, 5, and 6 inch wafers. The Chinese government is seeking business partners from industrialized nations for transferring 0.5 μm (and smaller) and 8 inch (and bigger) semiconductor manufacturing technology.

Despite rapid increases in production capacity and technological sophistication, China's IC production in 1995 and 1996 was unable to meet the domestic demand. Imports accounted for two-thirds to four-fifths of China's needed ICs. LSI and VLSI products are almost entirely dependent on imports. The Chinese government continues to make great efforts to increase the local production of ICs and reduce the dependence on importing IC products.

Tables 5.1 and 5.2 display China's remarkable growth in domestic IC demand and production, as well as the status of IC import and export. Figure 5.1 breaks down China's semiconductor market by type of application. Color TV was the largest user of IC products; however, the percentage has dropped from 33 percent in 1992 to 27 percent in 1995.

Table 5.1

Total IC Demand and Production in China (million units)

Year	Total IC Demand[1]	China's Total IC Output[2]	IC Output Need Met Domestically(%)	Total Output of IC Products & Chips[3]
1993	850	178	21	180
1994	1,000	245	24	250
1995	1,500	515	33	786
1996	2,200	758	34	1,149
1997*	2,800	800	29	N/A
2000 (forecasted)	4,200	2,000	~50	2,500

* Estimated

1. SEMI® estimates, Weng 1996 and CEInet China IT Market Report 1997. 2. China Electronics Industry Yearbook 1997, p.146. 3. CEInet China IT Market Report 1997.

Table 5.2

China's IC* Import and Export Markets

Year	Import Units (billion)	Export Units (billion)	Import Value ($)	Export Value ($)
1995	5.9	0.9	2.2	0.37
1996	6.9	1.3	2.7	0.60
1997 (forecast)	8.0	1.5	3.1	0.75

* Including microelectronics components

Source: *China Electronics Industry Yearbook 1997*, p. 147.

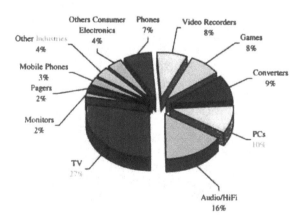

Figure 5.1 China's semiconductor market, by use, 1995.
Source: Howell et al. 1995.

5.1.4   IC Design

Until 1995 IC design was not done within the industry in China, but rather in universities and research institutes. One of the premier institutes is the China IC Design Center, which was founded in 1986. The focus is primarily on microelectronics, memory, IC cards, and MPEG-2 decoders; using the Panda 2000 VLSI CAD system, which provides a series of tools for high-level design, layout, verification, and layout migration [Liu 1998a]. Fudan University in Shanghai is another one of the many institutions that actively support China's IC industry in design and testing [IEEE 1995].

As of 1998, there were more than one hundred design houses and about one thousand experienced engineers. According to a survey by the China IC-CAD Federation, about 26 percent of these are independent IC design centers: 33 percent focus on connection technologies with universities and institutes, 12 percent belong to semiconductor facilities, 14 percent are subsidiaries of system assemblers, and the remaining 15 percent are new IC design houses created by foreign or Taiwanese investors [Liu 1998a]. For example, as early as 1995, Shougang NEC in Beijing was sending its engineers to design ICs at NEC's facilities in Japan. Shanghai Belling has about thirty engineers designing ICs in-house, with a goal of fifteen to twenty new designs each year.

As market demand for IC products has increased, China's IC design industry is gathering government support and foundries with growing technical expertise, are foreign investment. For example, under the government-sponsored Project 909 plan, a joint venture was established between Hua Hong Semiconductors and NEC to operate a foundry facility in Shanghai.

5.1.5   IC Technology and Product Development Status

Despite the rapid increase in production and technology, China's IC industry still lags far behind the technological levels of advanced countries. The Chinese government is determined to push forward China's IC design and manufacturing capabilities through collaboration with global technology partners. Table 5.3 shows the status of IC technology development in China and several major technological cooperative ventures between foreign multinational corporations and domestic semiconductor companies in 1998.

In addition to cooperation with foreign partners to develop and transfer technologies, several organizations and research departments have been established over the years by government agencies and universities to advance domestic IC technology development and manufacturing capability. Table 5.4 presents seven of such research institutes and the corresponding IC technologies that are under development. In order to narrow the technological gap between China and industrial nations, China must extend its efforts to develop indigenous advanced design and manufacturing capabilities.

Table 5.3
IC Manufacturing Technology Status in China

Company	Foreign Partner	Location/Chinese Partner	Product Sector	Technology 1996-1997	Planned Fabs for New Technology
ASMC (Worldwide)	Philips Semiconductors & Northern Telecom (Nortel)	Shanghai	Wafer Foundry	5", 2-3μm Bipolar / 6", 0.8μm CMOS / 5", 2-3μm MOS	6", 0.8μm CMOS / 8", 0.5μm MOS (1998)
Shanghai Belling	Alcatel (Belgium)	Shanghai/Shanghai Instrument-Electronic Stock Holding Group	Telecom	4", 1.2-2μm MOS / 4", 3μm BiCMOS	8", 0.5μm MOS (2000)
Shougang-NEC	NEC (Japan)	Beijing/ Capital Iron & Steel Co.	4 Mbit DRAM	6", 0.8μm CMOS / 6", 1.2μm MOS	6", 0.5μm MOS
Huajing	-	Wuxi/ Jiangsu Province	Consumer	4'-5", 2-3μm Bipolar / 5", 2-3μm CMOS / 5", 1-1.5μm MOS	6", 0.8-1μm CMOS (1998)
Huayue	-	Shaoxing/ Zhejiang Province	Consumer	3", 5μm Bipolar / 4", 3-5μm Bipolar	5", 1.5-2μm Bipolar / 6", 1.5-2μm CMOS
Shanghai Hua Hong NEC	NEC	Shanghai/Hua Hong Electronics	Memory/Logic		8", 0.35μm CMOS

Source: *IEEE Spectrum*, December 1995; D. Greene 1996; and Tsuda 1997

Table 5.4
IC Technology Development in China

Research Institution	IC Manufacturing Technologies R&D
Central Research Institute, Hua Jing Electronics Group Corp.	1μm (CSC245), 1M MASK ROM, DSP high-speed digital signal processing circuit, 256K SRAM
Microelectronics Research Center, Tsinghua University	1~1.5μm VLSI, 1M ROM
47th Institute (Shenyang city), Ministry of Electronics Industry (MEI)	1.5μm
214 Institute, Weapon Industry Corp.	3μm ASIC
Microelectronics Research Center, Academy of Science of China (ASC)	0.8~1.2μm
ASC Shanghai Metallurgy Research Center	1μm
Microelectronics Technology Research and Training Center, Aerospace Industry Corp.	ASIC, GJB protocol microprocessing circuits

Source: Summarized from Weng 1996

5.2   DOMESTIC SEMICONDUCTOR MANUFACTURERS

Of China's current total of 330 semiconductor plants, 25 produce ICs and the rest discrete devices. The five largest and most advanced companies are Shanghai Belling Microelectronics Manufacturing Co. Ltd. and the Advanced Semiconductor Manufacturing Corp. (ASMC), Shanghai; Hua Yue Microelectronics Co. Ltd., Shaoxing; Huajing Electronics Group Co., Wuxi; and Shougang NEC, Beijing. Most of China's major semiconductor facilities are partly or wholly foreign-owned by companies such as NEC, Matsushita, SGS-Thomson, Philips, Northern Telecommunications, Samsung, Motorola, Harris, and Intel. China's dominant semiconductor technology at present is 1.2 μm, well behind the 0.35 μm of the West and Japan. Many advanced equipment makers are selling China their older machines for 1 to 1.5 μm specifications. Chinese facilities are making large deals to acquire foreign semiconductor manufacturing equipment, process software, and know-how for both common and state-of-the-art technologies. Although China's government is encouraging foreign investment as a means to hasten technology advancement, it is working on major projects to lessen its dependence on foreign chip suppliers. Several of China's largest and most advanced IC manufacturers are described below, in order of their founding.

5.2.1   Hua Jing Electronics Group Corporation

Hua Jing Electronics Group is the largest of China's state-owned semiconductor plants and a subsidiary of China Electronics Corporation (CEC), a holding company that supports China's electronics industry. Hua Jing is located about 150 km northwest of Shanghai in the city of Wuxi in Jiangsu Province. It began with China's purchase in the early 1980s of a turnkey, second-hand, 3 inch line from the United States. Hua Jing now has the strongest technological capability of any indigenous semiconductor enterprise in China. Its principal business is the development and manufacture of discrete devices and both bipolar and CMOS integrated circuits [IEEE 1995], primarily for television sets and audio equipment. As of early 1997, Hua Jing had started production of 6 inch CMOS wafers with 0.8 to 1 μm design rules, with a monthly capacity of ten thousand wafers [Tsuda 1997]. In addition to its IC manufacturing lines, Hua Jing produces almost all of its own silicon wafers and maintains an R&D center that develops and tests new process technology.

Hua Jing relies heavily on the international semiconductor community for its technology support. The technology for 125 mm-diameter wafers was obtained from Siemens AG of Germany [IEEE 1995]. Support for bipolar technology comes from Toshiba Corporation (Japan), and manufacturing software comes from Promis Systems (Canada), including the ~$0.5 million purchase in September 1996 of Promis' Manufacturing Executive System software. In January 1998 Hua Jing completed a technology transfer from Lucent Technologies Microelectronics Group, which began in 1993 with an agreement between the State Council and Lucent, for worker training, processing technology, and related design tools for a 150 mm, 0.9 μm, single poly double metal complementary metal oxide semiconductor wafer [China Vista 1998]. Also, Intel has licensed Hua Jing as one of its testing and packaging partners for selected chips [Intel 1998].

Chinese authorities clearly intend for Hua Jing to be a "national champion" in the development of the country's semiconductor industry [Howell et al. 1995]. The Wuxi Hua Jing Expansion Project to upgrade the semiconductor manufacturing facilities and construct the IC research center was one of a handful of leading national projects economy-wide considered to be essential to national development. The center has a Class 10 cleanroom that meets requirements for 0.8 μm ICs as well as those with 2 to 3 μm design rules.

5.2.2   Shanghai Belling Microelectronics Manufacturing Company
Shanghai Belling was the first joint venture semiconductor manufacturing company in China, founded in September 1988 by Shanghai Electronics and Operation Instruments Holding Company, Radio Factory 14, and Shanghai Bell Telephone Equipment Manufacturing Company (itself a joint venture with Alcatel Bell, the Belgium branch of Alcatel). The total investment was $82.4 million [Belling 1996]. Shanghai Belling is located in Cao He Jing, a well-established, high-technology development zone in southwestern Shanghai. It has over five hundred employees, about 40 percent of whom are engineers and technicians. The company uses a Western-style, team-oriented management structure rather than a Chinese-style structure. Most of Shanghai Belling's revenue comes from ICs made for use in the private branch exchanges of Shanghai Bell Telephone, the first switch-maker in China to use locally made circuits [IEEE 1995]. The remainder comes from sales of micro-controllers and memory chips for use in such consumer

products as appliances and remote control units for television sets and compact disk players. Shanghai Belling manages its quality system according to ISO 9001 standards.

As of 1996, Shanghai Belling had one fabrication line, a 2000 m^2 Class 10 facility that annually produces over 160,000 1.2 μm bipolar and CMOS ICs on 100 mm (4 inch) wafers [Belling 1996]. Belling plans to upgrade its IC manufacturing technology to 6 inch, 0.8 μm by 1998 [Koo 1997.] and make ICs with 0.5 to 0.8 μm feature sizes on 200 mm (8 inch) wafers in the near future [IEEE 1995].

5.2.3   Advanced Semiconductor Manufacturing Corp. of Shanghai

Advanced Semiconductor Manufacturing Corporation of Shanghai (ASMC) was established in 1988 as a joint venture between Philips NV of the Netherlands and a group of Chinese investors. Northern Telecom, Ltd., (Nortel) of Canada joined the partnership in 1994, and its technology is the basis for ASMC's second line. ASMC, like Shanghai Belling, is situated in the Cao-He-Jing high technology park in southwestern Shanghai, which offers tax-free exports. The company draws most of its technical staff from Fudan and Jiaotong Universities in Shanghai, two of China's premier universities. It employs over 450 people.

ASMC has two wafer fabrication lines. Fab I is a Class 10 fab able to produce monthly 23,000 5 inch bipolar CMOS and power MOS wafers with linewidths of up to 2 μm. In 1997 ASMC produced more than 200,000 5 inch wafers and aimed to produce about 250,000 in 1998. Fab II, which became operational in March 1997, is a Class 1 wafer fab able to produce monthly 16,000 6 inch CMOS wafers of up to 0.8 μm. ASMC produced roughly 30,000 wafers in 1997 and should produce over 50,000 in 1998 [ASMC 1998].

ASMC's process portfolio includes 3 μm analog, bipolar, single/double metal CMOS wafers up to 60V for TV and telephone applications; 3 μm low-voltage (1.5-9V) CMOS with EEPROM option for telecom and consumer applications; and 1 μm single poly/double metal and double poly/double metal CMOS [ASMC 1998]. Philips and two U.S. companies have been using the bulk of the factory's output.

ASMC started strictly as a foundry and did not sell ICs of its own design, servicing IC manufacturers whose own fabrication lines were at capacity, as well as so-called "fab-less" semiconductor companies. Although this is still its primary market, it is working on coordinating design, assembly, and testing of its products. As proof of ASMC's growing status and capability in semiconductor manufacturing, it achieved ISO 9002 certification in January 1995, QS-9000 certification in February 1997, and plans to achieve ISO 14001 certification in August 1998.

5.2.4   Hua Yue Microelectronics Corporation

Hua Yue is another state-owned semiconductor business owned and controlled by CEC, but is less competitive. It sells its products on the merchant market and so is unlike either ASMC, which is a pure foundry, or Shanghai Belling, which sells most of its products to a single partner [IEEE 1995]. The company, located in the city of Shaoxing, manufactures bipolar ICs for television sets and telephones. In 1995-1996/7, Hua Yue started 15 to 17 thousand wafers with 3 to 5 μm feature sizes per month, of which seven thousand were 100 mm in diameter and the remainder were 75 mm [IEEE 1995, Koo 1997]. The company is expanding its capabilities to include 125 to 150 mm lines with 1.2 to 2 μm design rules that will enable it to produce ~50 million ICs per year [Howell et al. 1995, Tsuda 1997]. It has benefitted less than Hua Jing from foreign technology, and has been searching unsuccessfully for a foreign partner [Koo 1997].

5.2.5   Shougang NEC Electronics Corporation

Shougang NEC, a joint venture of Japan's NEC Corporation and the Capital Iron and Steel Company of Beijing, was founded in Beijing in 1991. This company designs, fabricates, assembles, and tests a variety of ICs, including linear devices, memories, microprocessors, gate arrays, and communications chips. A new manufacturing plant, office, and dormitory building were completed in October 1993, assembly operations started in 1994, and wafer fabrication began in March 1995. The company employs over eight hundred persons, and most of the engineers are trained in Japan. As of 1996, the facility was assembling a maximum of four million 16 Mbit DRAM units a month, and processing three to four thousand 6 inch, 1.2 μm wafers a month for 4 Mbit DRAMs, MCUs, and other ICs, corresponding to forty-seven million units, well above original projections [Tsuda 1997]. Ramp-up to five to eight thousand wafers a month was achieved in December 1996 as planned [Lammers 1997]. NEC provides production and management technology, including advanced LSI circuit diffusion and packaging production lines and testing equipment. The Chinese share in the joint venture started at 60 percent and has decreased to 49 percent; NEC's stake in the joint venture, which has risen from 40 percent to 51 percent. The total first-phase investment was about $240 million for 4-bit micro-controllers as well as the 4 Mbit and 16 Mbit assembly operations. A further investment of over $100 million was made for the production of 0.5 μm devices for 16 Mbit DRAMs [Tsuda 1997, Lammers 1997].

In the third quarter of 1998, Shougang NEC introduced the first 64 Mbit DRAMs [China Economic Information Net IT News, 19 August 1998]. This indicates that China has made tremendous progress over the years to improve and upgrade its products and technologies in order to reach the world's advanced standards.

5.2.6   Planned Fabs

China is pushing plans for construction of several submicron fabrication facilities with supports from government. This section presents some of the major efforts under way in China.

5.2.6.1   Motorola

Motorola is building a wholly-owned submicron facility in Xianing, south of Tianjin city. Construction began in 1995, and has an estimated total cost of $1.2 billion. Under the plan, an 8 inch wafer-processing line is being built to begin processing devices with 0.8 µm technology in 1998 and 0.5 µm technology in 1999. Monthly capacity will eventually reach about twenty thousand wafers per month. Major applications include telecommunications and automobile electronics [Tsuda 1997].

In May 1998, Motorola decided to double the size of the Tianjin wafer-processing facility by spending $2.6 billion to turn the site into a "superfab" and a linchpin in its Asian operations. The Tianjin manufacturing complex will contain both high-volume, front-end, wafer-fab lines and advanced back-end chip-assembly operations. The second phase of the production plan calls for a 0.35 µm fab line to come on line in 2000, doubling the silicon-processing capability of the site [Robertson 1998b].

Motorola's other major investment, recently approved by the Chinese government, is the $225 million Leshan-Phoenix Semiconductor Company located in southwest China's Sichuan Province. Upon completion, Leshan-Phoenix will be capable of producing 7.5 billion units of semiconductor elements and devices, one billion ICs, and 250 million power transistors annually [China Economic Information Net IT News, 27 August 1998].

5.2.6.2   Shanghai Hua Hong NEC Electronics Co., Ltd./Project 909

As a vital step in the Chinese government's ninth FYP Project 909 to root an advanced semiconductor industry in China, it selected NEC in October 1996 as a joint venture partner with Chinese partners to design, manufacture, and market memory and logic semiconductors using 0.5 to 0.35 µm process technologies [NEC 1997]. The original Chinese investors were China Electronics Company (CEC), Shanghai Jiushi Company, and Shanghai Instruments Group. The Hua Hong plant is 93,700 square meters in area, of which 62,000 square meters are for plant construction, and 5,000 square meters are for the clean processing. When the fab achieves maximum capacity in September 1999, it will produce twenty thousand 8 inch wafers per month for memory and logic ICs. The fab is being built in Pudong, Shanghai, for a total investment of about $1 billion. NEC's share totals 28.6 percent, and Shanghai Hua Hong Microelectronics Co., Ltd.'s share is 71.4 percent. Employees are expected to number five hundred in 1999 and seven hundred by 2001. At first, the company will sell its products only in China, and later, expand sales into Asia and other areas of the world. As part of NEC's partnership agreement, it is establishing a working partnership with MEI through which NEC can enter into other business in China, but must also negotiate financial and management arrangements such as product mix, where disagreements have existed between the Chinese and Japanese investors.

5.2.6.3 Beijing Hua Hong NEC IC Design Company

A new joint venture between NEC and Beijing Hua Hong Integrated Circuit Design Company was established in February 1998 to design and sell semiconductors. The Beijing Hua Hong NEC IC Design Company aims to provide Project 909 with two hundred kinds of ICs and twenty thousand units of 8 inch silicon chips by 2001 [China Economic Information Net IT News, 11 September 1998]. It will focus on designing microcomputers, ASICs, IC cards, and other semiconductor products for use in applications in the areas of digital video and still cameras, consumer electronics, and mobile communications equipment. In addition, system on chip (SOC) devices will also be designed by the joint venture company. Devices designed by the company will be produced at Shougan NEC or Shanghai Hua Hone NEC. Overall investment in the company will be $30 million, 60 percent owned by NEC and its affiliates, including Shougang NEC [Williams 1998].

5.2.6.4 Others

Fujitsu is among other foreign semiconductor firms planning new fabs in China. The Fujitsu joint venture, Nantong Fujitsu Microelectronics Co., is a $10 million undertaking to assemble and test semiconductors. When it begins operations in 1998 it will initially produce about ten million units per month, including micro-controllers and linear ICs. Fujitsu's interest will be 40 percent.

Intel dedicated its first manufacturing facility in China in May 1998. It is a $198 million test and assembly plant for manufacturing flash memory chips for the worldwide market. In addition, Intel plans to build a $50 million information technology research center aimed at Internet-related issues and other applications with specific relevance to China, including speech recognition [*Semiconductor Business News, 7 May 1998*].

In cooperation with Canadian Northern Telecom, the Shanghai Xianjin Semiconductor Company established a joint venture for processing 2 to 3 μm, 5-inch bipolar IC products. The output in 1994 amounted to fifty six million ICs. The next step is to build an assembly line for producing 0.8 6 inch ICs and a special purpose circuit for processing program-controlled exchanges and other functions [China IT Market Report 1997].

One of the largest IC manufacturing projects in China recently is the Mitsubishi-Stone Company, a joint venture of Japan's Mitsubishi Electric Corporation and Mitsui Co., Ltd., and the Beijing Stone Group Company. The company has a registered capital of $35 million, of which the Japanese side holds 70 percent and the Chinese side 30 percent of the shares. The project is expected to produce 210 million ICs, ranging from 0.5 to 0.35 μm. The first phase of this project went into production in the fourth quarter of 1998 [China Economic Information Net IT News, 16 October, 1998].

Finally, Motorola launched an advanced materials joint research program to investigate fundamental properties of ferroelectric thin-film materials. This class of materials has potential application for advanced non-volatile memory for cellular phones and smart cards. The program will draw upon the technology strength of the National Lab of Solid State Microstructures at Nanjing University and the technology application capability of Motorola [*Semiconductor Business News*, 2 April, 1998].

Table 5.5 summarizes the technology, production capability, and background of the major semiconductor companies operating in China. It shows that China's IC semiconductor sector still has a long way to go before matching the technological and production levels of technologically advanced countries.

5.3   ROLE OF FOREIGN COMPANIES

As the foregoing indicates, there is considerable foreign involvement in China's semiconductor industry. In general, China welcomes investments of foreign companies that upgrade the overall technology level of the IC market, and the government often seeks a foreign technology transfer partner for the newest IC manufacturing technologies in projects such as that described above. However, China seeks to maintain control over the direction and decision-making of companies operating in China, and this can at times be a problem for foreign companies doing business there. Other problems for foreign companies include lack of adequate protection for intellectual property and inconsistency between different branches of the government. These are areas that China is addressing. For U.S. companies there is the additional problem of working within U.S. laws that govern export of "dual use" products and equipment that could be turned to military purposes considered detrimental to U.S. security interests.

For a decade now, companies from the United States, Europe, and Asia have been building and equipping factories in China, training engineers and operators, and co-managing manufacturing operations, despite less than ideal conditions. Companies active in China's semiconductor industry, drawn by the low-wage labor force as well as by the huge potential Chinese market, include AT&T, IBM, Intel, Fujitsu, Motorola, Mitsubishi, National Semiconductor, NEC, Philips, Rockwell, Siemens, Texas Instruments, and Toshiba — all intensely competitive with one another. In addition, China is separately purchasing processing equipment from the United States, Japan, and Europe to help expand production capacity, particularly in 0.8 to 1.0 µm and 6 inch technology, and there are opportunities for equipment suppliers offering 0.5 µm (and smaller) and 8 inch technology, as well. The Motorola and Hua Hong-NEC plants will give China its most advanced chipmaking

Table 5.5

Major Semiconductor Companies in China

Company	Ownership	Technology	Wafer Starts/ Month	Background
Hua Jing Electronic Group	China Electronics Corp. (CEC)	3", 4", 5", 2-3 μm bipolar and MOS;6", 0.8-1.0 μm CMOS technology transfer from Lucent	25,000 for bipolar and MOS devices in production; 6,000 for CMOS emerging from startup	China's semi-conductor leader; recipient of Lucent technology transfer as part of state-funded project 908. In 1995, produced 139 million discretes worth ~$34.3 million and 84.6 million ICs worth ~$68.9 million
Shanghai Belling Micro-electronics	Belgium Tel., Alcatel, Shanghai Bell, Shanghai Municipal Government	4",5",1.2-8 μm MOS; will upgrade to 6" 0.8 μm in 1998	12,500	Established 1988, 5 years after Shanghai Bell, China's dominant manufacturer of telephone switch equipment. In 1995, the venture produced 28 million ICs worth ~$51.3 million
ASMC Worldwide	Shanghai Electronics Bureau (SEB), Philips and Northern Telecom	5", 2-3 μm bipolar/CMOS; 6", 0.8-1.0 μm CMOS	23,000 for 5";16,000 for 6"	SEB and Philips JV established 1988; later joined by Nortel. As a foundry, did not report on ICs produced but recognized sales of ~$29 million in 1995 for about 110,000 wafers processed.
Hua Yue Micro-Electronics Corp.	China Electronics Corporation	1.5-2.0 μm Bipolar MOS	15,000-17,000 at 3-5 μm	In operation since 1986 with 5 μm 3"; 4" in 1992. Manufactures bipolar ICs for TVs and telephones
Shougang-NEC Electronics	60/40. Capital Iron & Steel/NEC	6", 1.2 μm MOS; linear, memory, micro-processor and gate arrays; 6", 0.5 μm DRAM	4,000 plus additional 4,000 at 0.5 μm DRAM underway	In operation since 1993; NEC has invested more than $300 million. Some tools transferred from NEC plants in Japan. Said to have achieved 90% yield on 0.5 μm line. In 1995, produced 39 million ICs worth $110 million
Shanghai Hua Hong NEC Electronics	Hua Hong Electronics and NEC (28.6%); Hua Hong is 50/50 JV of CEC and two Shanghai entities.	8", 0.35 μm CMOS	20,000	As of 1997, construction under way, but investors have yet to agree on product mix for this $1 billion fab. NEC has operating control and must meet certain performance milestones.
Motorola	Motorola	8", 0.5-0.35 μm MOS; devices for pagers and cellular phones	15,000-20,000	Motorola plans to invest $2.6 billion to turn the Tiajin facility into a "superfab" and a linchpin in its Asian operations.

Source: Weng 1996, Koo 1997, Tsuda 1997, ASMC 1998, and Robertson 1998

capabilities to date. The facilities will begin with 8 inch wafer, 0.5 μm processing. Both fabs hope to produce 0.35 μm wafers in the year 2000.

5.4   IC INDUSTRY ASSESSMENT

In 1998, the Chinese IC industry could be characterized by the following features:

- inadequate capacity and technology. China satisfied only 20 to 30% of domestic need in 1996-97.
- strong competition. China must compete against Western and Asia-Pacific companies equipped with much better technologies and financial resources.
- poor infrastructure. China's IC infrastructure is poor and immature. In addition, lack of peripheral industries (e.g., equipment manufacturing, assembly, and testing) is a major concern for future growth.
- inadequate design and manufacturing capabilities and low productivity.
- inadequate research and development. Most technologies are acquired from foreign countries.

The current U.S. trade policy toward China is to deny export and technology licenses for fabrication equipment that can produce ICs using below 0.5 μm process technology [*EE Times*, 27 April 1998]. The U.S. position is that export controls are needed to prevent the Chinese from making high-tech ICs for missile and nuclear weapon technologies. As a result, for China to become a significant player in the global IC market, it must exten the scope of international cooperation and speed up the current progress in indigenous technology development. The increasing domestic demand for computer and telecommunications equipment and devices will benefit the future growth of China's IC industry.

5.5   DISCRETE SEMICONDUCTORS

In 1996 China's discrete semiconductor industry produced almost eleven billion units, mostly diodes and thyratrons (Table 5.6). However, due to intensive global competition, the amount of production and total sales decreased by 10 percent in 1995 and 5.6 percent in 1996. Table 5.7 shows that the growth of exports in discrete semiconductors suffered tremendously in 1996, while the amount of imports increased slightly, about 2 percent.

While many multinational semiconductor firms are decreasing production of low-end products in favor of advanced devices, China's ninth FYP supports the production of low-end products as well as improving the technological sophistication and output of discrete devices.

Table 5.6

Discrete Semiconductor Production and Sales in China (million units)

Category	1995		1996	
	Production	Sales	Production	Sales
Diodes	9,755.8	9,086.5	8,495.0	8,341.0
Thyratrons	2,385.5	1,795.5	2,348.2	2,191.0
Other	245.0	326.2	22.6	26.8
Total	12,386.3	11,208.2	10,865.8	10,558.8

Source: *China Electronics Yearbook* 1997, p.142.

Table 5.7

Discrete Semiconductor Export-Import Status in China

Category	Unit	1995		1996	
		Imports	Exports	Imports	Exports
Quantity	Million	19,868	22,996	23,041	23,822
Amount	$million	673	537	688	323

Source: *China Electronics Yearbook*, 1997, p.144.

5.6   PACKAGING AND ASSEMBLY

ICs are first designed, then produced and tested, and then packaged and assembled; in China the most robust of these three main areas is packaging [IEEE 1995]. Electronic packaging in China has developed rapidly since the 1980s, from labor-intensive simple assembly to highly specialized and hybrid packaging.   This has involved critical involvement of the Chinese government, the Chinese Electronics Packaging Society, and foreign electronics firms, in addition to the efforts of highly motivated domestic firms.

5.6.1   Assembly and Packaging Technology

This section discusses the recent developments of China's electronic assembly and packaging technology, as well as the market demand condition for China's printed circuit boards.

5.6.1.1   Printed Circuit Boards

As labor costs in Hong Kong increased in the early to mid-1980s, contract manufacturers there began to move their manufacturing and packaging facilities into China, particularly for those labor-intensive processes for which inexpensive manual labor is most cost-effective. Wong's Electronics was the first Hong Kong contract manufacturer to install surface-mount assembly equipment in China in this period [Boulton 1997]. Other foreign manufacturers followed suit. Companies from Taiwan, Japan, the United States, and Europe all have established assembly plants in China. Until recently, most printed circuit board (PCB) assembly in China of through-hole components has been done manually, taking advantage of the low average wage of Chinese workers. China's PCB production was developed mostly to serve consumer electronics markets, including those for televisions, radios, and tape recorders.

Most PCBs produced in China for Hong Kong and other contract manufacturers in the 1980s and early 1990s were simple, single-sided boards. Only 20 percent of boards were more sophisticated, for use in the computer and telecommunication industries [Howell et al. 1995], and complex boards were merely assembled from complete or semi-knockdown kits. With the increasing sophistication of electronics products generally, the amount of labor-intensive through-hole assembly is declining in China. Today's electronics products require surface-mount devices (SMD) to achieve smaller product size and lower weight. PCB assembly of through-hole components is now reserved for electronic toys and products that have no miniaturization requirements.

Recent examples of international cooperation in PCB production include CPC Inc. of Randolph, Massachusettes ($8 million revenue in 1998) and Dalian Pacific Mult-Layer PCB Co. Ltd. in Dalian, China. The partnership will allow CPC, a niche manufacturer of instrument controls and medical instrumentation, to expand its prototype and early production efforts with high-volume production and a greater number of customers [Dunn 1998].

China's PCB production in 1992 was valued at $394 million; production in 1997 was expected to reach $735 million. During the same period, China's consumption of PCBs was estimated to grow 69 percent, with the largest growth in flexible and multi-layer PCBs [Howell et al. 1995]. Table 5.8 presents forecasts of China's printed circuit board demand by type of application through 2005.

5.6.1.2 Types of Packages Manufactured in China

There are four categories of electronic packages produced in China: plastic, high density, metal, and ceramic.

Plastic packages. In 1996, plastic packaging constituted over 95 percent of the total packaging in China. CEPS claims that there are up to fifteen billion packages being made in China. With the development of the semiconductor industry of China, CEPS estimates that by 2000 the country's plastic packaging needs will increase to thirty billion packages.

High-density packages. Chinese companies have developed various kinds of high-density packages, including 209- to 300-pin PGAs (pin-grid array) and BGAs (ball-grid array), and 44- to 200-series PQFPs (plastic quad flat pack) and MCMs (multi-chip modules). Chinese companies have also developed high-density lead frames and high-density sockets. Market demand is estimated by the Chinese Electronic Packaging Society (CEPS) to be above two million in the next five years.

Table 5.8
Printed Circuit Demand in China and Hong Kong, by Application

	Forecast Electronic Equipment Production ($ millions)				Forecast PCB Demand ($ millions)			
	1994	1997	2000	2005	1994	1997	2000	2005
AV Equipment	6,729	10,914	13,355	15,934	801	494	561	598
Communications Equipment	4,766	4,731	5,822	7,743	0	172	213	287
Computers	4,184	6,591	11,676	22,918	0	143	256	508
Office Equipment	1,217	1,496	1,838	2,253	0	61	76	93
Electrical Instruments	572	653	858	1,335	0	71	89	131
Other Applications	99	124	154	205	0	14	17	21

Source: Japan Printed Circuit Association [Hatori 1996]

Metal packages. Low-cost metal packages are made for hybrid ICs, optoelectronic devices, and selected discrete devices. The M16-M100 series is a main product. The Chinese demand for metal packages is 2 to 4 billion per year. In China, metal-ceramic packages have been developed for microwave and millimeter wave discrete devices, and MMIC. Directions for development are mainly towards high-frequency, high-power, and low-noise devices. Most of these are determined by the development needs of the Chinese military.

Ceramic packages. Conventional ceramic packages in China include the D14-D16 and F14-F24 series. The demand for ceramic packages per year is estimated by CEPS to be above five million.

5.6.2   Processing Technologies

Various processing technologies, such as chip-on-board, hybrid microcircuit packages, and microwave and millimeter wave packaging, have been adopted and improved since 1980s in China's assembly and packaging industry.

5.6.2.1   Chip-on-board Technology

In the chip-on-board (COB) process, bare chips, without the packaging, are mounted directly onto the printed circuit board, thereby reducing the dimension and weight of the final electronic assembly. NamTai Electronic (Shenzhen) Co., Ltd., of China, in conjunction with its Hong Kong affiliates, is developing COB technology.

NamTai's current design rule for track width and space design of the PCB pattern is usually larger than 0.15/0.15 mm. However, it can mass-produce PCB copper patterning, which is different from electroplating. This process does not require special plating of lead patterns and can produce high-density boards. A staggered pattern is used for high pin-count, fine-count and fine-pitch ICs. The PCB materials used in COB are mainly worn glass-fiber in epoxy (FR-4, CEM-3). Products with multichip and large PCB areas are being produced with rotary head-type aluminum wire bonding machines.

5.6.2.2 Hybrid Microcircuit Packages

The annual demand in China for hybrid microcircuit packages was estimated at approximately seven million pieces in 1996, while the domestic production of such packages was about five hundred thousand. As a result, a large number of hybrid packages had to be imported from abroad that, in turn, contributed to higher production costs.

Hybrid technology is one of the ways to achieve miniaturization of complex electronic systems. The conventional hybrid package widely used in China is the Kovar package, and materials for package pins or leads include FeNi alloys such as 4J29, 4J42, and 4J52. Metal components like seal rings and heat sinks are attached through brazing. Gold, tin, nickel and various combinations of plating are available for specific applications. The East China Microelectronics Research Institute is one of the major organizations specializing in hybrid packaging technology in China.

A variety of hybrid packages has been developed and produced for both commercial and captive applications. Wuhan Radio Devices Factory, a manufacturer specializing in semiconductor components and devices packaging, has developed a series of metal packages with pin numbers ranging from ten to sixty-four. Packages of leadless chip carriers (LCC) developed by Yixing Electronic Device Factory have been used in many hybrid products.

Although many package manufacturers have improved their processing capabilities after installing new equipment such as tape casters or matching and plating facilities, the overall technological level and production capability still cannot satisfy the domestic demand for packaging and assembly. In addition, outmoded equipment and manufacturing processes continue to be a major problem in China.

5.6.2.3 Microwave and Millimeter Wave Packaging

In the area of microwave and millimeter wave packaging, the assembly process consists of discrete devices, MMICs, MCMs, and other components. The major characteristics of microwave and millimeter wave packaging are higher performance and more accurate electronic and mechanical parameters. It is suitable for low-frequency digital circuit packaging and assembly. As frequency increases, the bandwidth is extended and power consumption rises, increasing the difficulty of design and fabrication, especially for systems operating in the millimeter wave region. Nanjing Electronic Device Institute has the ability to develop and produce a small quantity of microwave and millimeter wave packages.

MMICs are primarily designed for military applications. However, since 1990, MMICs have been manufactured for civilian use, such as mobile communications, satellite communications, and traffic management. In 1994, more than fifty types of system components were produced using GaAs

MMICs, including receivers and control subsystems, broadband amplifiers, and voltage-tuned attenuators. In addition, microwave circuits have been developed for hybrid, monolithic integrated, multi-chip micro assembly, and micro system integrated devices.

Major goals for improving China's microwave and millimeter wave packaging technology in the coming years are as follows:

- Increase power and efficiency of the packaging cases of microwave GaAs power devices.
- Ensure higher power output of broadband in the case of microwave tin powder devices packaging.
- Achieve commercial application in GaAs MMIC packaging.
- Increase the bandwidth of MMICs close to that of hybrid integrated circuits.
- Emphasize research on the application of NF ICs and 8mm devices.

5.6.3 China's Electronic Packaging Businesses

The electronic packaging business in China has been growing since the 1980s, due to increasing demand for IC products around the world. This section introduces the Chinese Electronic Packaging Society and briefly discusses the activities of foreign and domestic packaging businesses in China.

5.6.3.1 The Chinese Electronic Packaging Society (CEPS)

The Chinese Electronic Packaging Society consists of more than sixty research institutes and companies and has nearly twenty thousand members. Major roles of CEPS include identifying and transferring advanced packaging technologies from industrial nations to China and coordinating information exchange activities with the industry. Other major activities of the CEPS include the following:

- monitoring members' electronic packaging activities, information exchanges, and technology development efforts;
- promoting international information exchanges and strengthening China's research and development capabilities;
- consulting on major national guiding principles and practices on electronic packaging;
- popularizing electronic packaging technology and diffusing technological know-how;
- safeguarding the legal rights and interests of Chinese scientists and engineers.

5.6.3.2 Major Foreign and Domestic Firms Involved in Packaging and Assembly

About fifteen manufacturers produce multi-layer PCBs in China, the majority of them producing boards of up to eight layers [Boulton 1997]. Shanghai Printronics Circuit Corp., a joint venture between Shanghai No. 20 Radio Factory and Australian Printronics Co., Ltd., are able to produce a maximum of fourteen layers. Shennan Circuit Corporation, a state-owned company located in Shenzhen, is capable of manufacturing circuit boards of from four to sixteen layers, as well as twenty-layer boards of 100 mm lines for applications in aerospace, computers, communication, and precision instruments. Bestman Electronics Co.,Ltd., another joint venture company, operates in Shenzhen and, is able to produce circuit boards of up to thirty layers [Boulton 1997].

In the area of electronics packaging and assembly, there are more than sixty companies and about fifty thousand support staff in China in the early 1990s. Shanghai Belling, Hua Jing, Hua Yue, and Shougang NEC are already performing their own IC packaging and assembly activities. Wuxi Huazhi Semiconductor Co., a joint venture between Toshiba and Hua Jing, is capable of packaging and testing bipolar IC products manufactured by Toshiba in Japan [IEEE 1995]. Other foreign multinational companies, such as Intel, Mitsubishi, and American Micro Devices, have also established IC packaging and assembly facilities in China.

Major projects to improve China's packaging and assembly technology under China's Ninth Five-Year National Development Plan include joint ventures with Mitsubishi in Beijing ($90 million), with SCG Thomason in Shenzhen ($78 million), and with Alphatec in Shanghai ($75 million). In addition, foreign companies that have established packaging and assembly facilities in China include Samsung and Advanced Micro Devices in Suzhou and Hyundai and Intel in Shanghai.

5.7 FUTURE DEVELOPMENT OF CHINA'S SEMICONDUCTOR INDUSTRY

China is one of the forty forces identified by the *EE Times* that will shape the global semiconductor industry [*EE Times*, 30 September 1998]. The power of China's 1.2 billion people, with their intrinsic respect for higher education and work ethics, cannot be ignored. The communist bureaucracy has held back China's advance as an electronics consumer and manufacturer for decades. However the giant is awakening, and China will reshape the landscape of the global electronics industry as it assumes its proper role on the world stage.

The major emphases of the development of China's semiconductor industry in the coming century are in the following areas:

- IC design (ICCAD) capability;
- wafer preparation and chip manufacturing, including polysilicon, crystal silicon preparation and process, doping process, pattern microfabrication, dielectric thin-film technology, metal thin-film technology, and clean room technique;
- IC assembly, testing, and reliability.

The 64 Mbit DRAMs made recently by the Shougang NEC Electronics Co., Ltd., a Sino-Japanese joint venture, indicate that China's manufacturing technology of super-large ICs is close to level of industrialized nations. By the year 2000, China will be able to achieve mass production of ICs by using 0.5 to 0.8 µm process technology. The more advanced 0.35 µm technology will be ready for implementation. Other areas targeted for major development efforts include 1Mb SRAM, 4 Mb ROM, IC cards containing EPROM, flash memory, 200k CMOS gate array, 300k CMOS standard cell, 50k BiCMOS gate array, and high-performance DSP. If the current pace of technological progress in China continues without disruption, it is expected that by the year 2005, 0.35 µm process technology will be implemented in mass production of ICs, and 0.15 to 0.25 µm technology will soon be ready for implementation. By the year 2005, China also expects to reach the technological level of 1M gate array, 250 K BiCMOS gate array, 4~16Mb SRAM, and 4~16Mb flash memory.

Chapter 6

CHINA'S INFORMATION ELECTRONICS INDUSTRY

China's overarching computer-related goal is to greatly enhance its capability to provide systems and equipment for the "informationization" of the national economy; to this end the government has targeted the computer and software industries, along with the semiconductor and telecommunications industries, for a concerted development thrust. In addition, China is determined to use its new strengths in computer-related fields to help grow its economy via exports, and it has a long-range commitment to international technical leadership in these areas, seeking to be one of the "Eastern giants" in computers and software along with Japan [Yang 1997].

Computerization of China's economy and its business practices is progressing rapidly, backed by government planning and financial and regulatory support, as well as by commercial and popular enthusiasm. Considering that the commitment to extending computer use throughout the country is only ten to fifteen years old and that this country of over 1.2 billion persons has had an essentially rural, traditional economy, the degree of progress is remarkable. In 1997, production from China's electronic-information industry reached $42.2 billion, a 23 percent increase from 1996. The figure is only slightly behind the $49 billion in electronic products that Taiwan is expected to produce in 1998 [Carroll 1998a].

Although China's information electronics industry has made tremendous progress in the 1990s, especially in ICs, software development, Chinese-language operating system software, and mobile communications, there is still some distance to go to catch up. In addition, there is a severe regional imbalance in computerization growth in China; it is rapid in certain cities and provinces (coastal areas especially), but has barely begun in other areas, particularly the more rural and inland areas. Likewise, adoption of computer technologies is significantly more rapid in large enterprises than in smaller ones. Widespread adoption of new technology is hampered by low average incomes and educational levels.

China's large, still barely tapped markets for a broad variety of computer and computer-related hardware, software, systems, and services are attractive to foreign and domestic firms alike and hence are highly competitive. According to International Data Corporation's estimates, from 1997 to 2000,

123

spending in China on information technology will be 19.4 percent, mostly on personal computers (44%). That is much higher than the current worldwide average of 10 to 11 percent [*PC Magazine*, 8 May 1998]. While China has facilitated the entry of myriad foreign firms into these markets, its own state-owned and private enterprises are working hard to capture and retain major market shares.

6.1 PERSONAL COMPUTERS

Chinese-built personal computers (PCs) are already "very advanced systems and very competitive with the multinationals," according to James W. Jarret, head of Intel's China operations. Chinese computer companies are working hard to bridge the gap with companies such as IBM, Compaq, and AST [Roberts and Einhorn 1997]. Their rapid improvement has less to do with Beijing's industrial polices than with the competitiveness of Legend Group, China Great Wall Computer Group, Beijing Founder Electronics, and other companies in Beijing's Haidian district (a hot electronics area of six thousand companies). These companies are manufacturing up-to-date products that are less expensive than foreign ones. For example, a Chinese-made PC with a 166-MHz Pentium processor, for example, cost around $1,200 in 1997, 20 percent cheaper than a similar IBM or Compaq PC [Roberts and Einhorn 1997]. Chinese computer vendors are competing fiercely to improve distribution and service to gain early market share.

6.1.1 National Development Goals

Major goals of China's Ninth Five-year National Development Plan to advance its electronic information industry include the following [China IT Market Report 1997, Simon 1996, Yang 1997]:

- developing massive parallel processing (MPP) "supercomputer" systems with computing speeds of fifty billion floating point operations per second; one MPP system is to be completed by the end of 1998;
- increasing the percentage of domestic components in Chinese-assembled computers, and generally increasing the nation's capacity to produce peripherals such as monitors, printers, floppy hard disk drives, and board cards;
- achieving a per capita national computer penetration rate of one percent, 20 percent among urban families;
- developing two to three domestic microcomputer manufacturers into enterprises with an annual production capacity of over $1 million;
- applying computer products and techniques to the renovation of traditional industries;
- promoting industrialization of multimedia computers and supporting products, such as high storage capacity equipment and high-definition displays;

Table 6.1
Production Goals for Selected Computer Products (million units)

Product Category	1997	2000
All microcomputers	1.5	4
Printers	3.5	2.5
Monitors	6	6
Floppy disk, hard disk, CD-ROM drives	25+	8
Intelligent cards (financial, non-financial)	10	45
Multifunction board cards	30	-
Automatic teller machines (ATMs)	-	0.03
Card readers	0.005	1

Data summarized from China IT Market Report, 1997.

- developing and enforcing uniform standards for China's computing environment via a PC production licensing system in order to help standardize the domestic microcomputer market and answer complaints about lack of service and intellectual property protection on clone PCs.

Table 6.1 summarizes and compares the production goals of 1997 and 2000 for major computer hardware equipment.

6.1.2 Brief History of Computer Use in China

China's computer industry started to develop in the late 1950s. In 1958, the first Chinese-made computer (a vacuum-tube computer called the 901) was manufactured at the prestigious University of Harbin, at the Institute of Military Engineering. During the 1950s and 1960s, China received foreign aid from the Soviet Union in many forms to assist in scientific research to develop computer technology. Responding to national security needs, in the 1960s and 1970s several Chinese-made computer series were developed, including the 100 and 150 series. These were installed in universities, military laboratories, and some important industrial conglomerates. On these computers China developed its own large computer systems, including a navy command system, missile-launching and satellite control systems, geological data analysis systems, production systems for oil fields, and similar operations [Zhang and Wang 1995].

In the 1980s, with the open door policy and economic reform, China reevaluated its strategy in the computer industry, and switched from research and development of large-scale mainframe computers to the development of personal computers. The Chinese PC industry has a history dating back only to 1985. In that year, the State Computer Industrial Administration selected a group of core technicians to form a scientific research task force. By June 1985 the task force had successfully developed a personal computer, Great Wall 0520CH, which was the first PC using Chinese character generation and display technology capable of processing information in Chinese.

The appearance of Great Wall 0520CH gave rise to the birth of China's PC industry. Soon after, Great Wall 0520CH began batch processing and gained a large share of the domestic market. As of 1996, many Great Wall 0520CH computers purchased by China's General Customs Administration were still operating normally, storing a large quantity of vouchers [Li 1996].

Since the 1980s, the rapid development and diffusion of information and communications technologies and products around the world has provided a huge growth opportunity for the Chinese PC industry. However, domestic production could not satisfy the increasing demands for personal computers in China. Hence, relevant government agencies have opened the door to foreign enterprises in the hope that imported foreign information products will stimulate the process of domestic PC technology development and accelerate the rate of domestic PC adoption. Notable name brands, such as Acer, AST, Compaq, IBM, and others, have set up sales offices or joint ventures in China. In the late 1980s, China increased its imports of large and midsize computers from U.S. and Japanese computer manufacturers such as IBM, DEC, Unisys, Fujitsu, Hitachi, and NEC.

Figure 6.1 compares the growth of China's GDP with the sales of personal computers from the early 1980s to 1994. As shown in Figure 6.1, since 1990 the rate of growth of domestic PC sales far exceeds the rate of growth of China's gross domestic product (GDP).

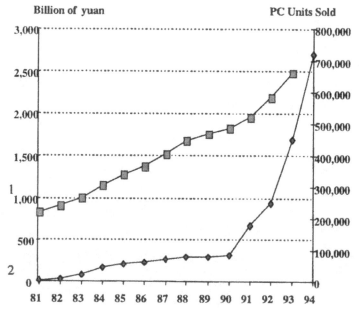

Figure 6.1 GDP growth (1) and domestic PC sales volume (2), 1981-1994
Source: Zhang and Wang 1995

6.1.3 China's Computer Market in the Mid-1990s

The growth and use of PCs in midsize and small businesses, and the acceptance and need for graphical user environments, such as Windows and related applications, are the two specific factors contributing to the boom in the Asian PCs market [Forbes 1997]. The Chinese computer market sold 2.1 million PCs in 1996 with a total value of $3 billion (24.6 billion yuan), an increase of 47 percent over 1995 [China IT Market Report 1997]. In 1997, China bought 3 million personal computers, an increase of 40 percent over the previous year. The monetary value of 1997 PC sales was $3.6 billion, up 17 percent from 1996, according to the data released from China's Ministry of Electronics Industry. Sales of PCs by Chinese companies accounted for 60 percent of the total domestic market [*EE Times*, January 19, 1998]. It is estimated that China's PC marketplace will grow at an even faster pace to more than 7 million units sold in the year 2000 [*Semiconductor Business News*, June 25, 1997].

Figure 6.2 shows the increase of PC sales volume in the 1990s. Compared with the rapid growth rate of annual PC sales, the slower rate of increase in monetary value was due at least in part to the fact that there was intense competition among local vendors. This led to price cuts that averaged 15 percent for companies like Legend and Founder, and to prices that were up to 60 percent less than those of foreign competitors for similar machines [*China Telecommunications Weekly*, 3 March 1997]. Thanks to such practices, domestic vendors were able to increase their market share from 20 to 26 percent between 1995 and 1996 [China IT Market Report 1997]. At the same time, foreign computer manufacturers, such as IBM, Compaq, AST, and NEC, set up their own factories in China, selling their products locally as well as exporting to overseas markets.

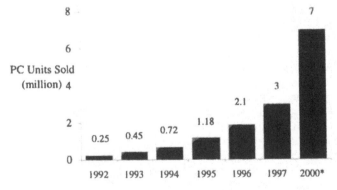

Figure 6.2 PC Sales Volume in China
Sources: China IT Market Report 1997 (1992-1996); *EE Times*, January 19, 1998 (1997); and *Semiconductor Business News*, June 25, 1997 (2000 *forecasted)

The 1996 company quality survey conducted by China's Ministry of the Electronics Industry (MEI) showed that only 40 percent of domestic PCs met the required quality standard set by the government. MEI has since pushed its new licensing standards and procedures. As a result, Chinese licensed vendors were able to increase market share at the expense of cloned PC products. By early 1997, MEI had granted 244 production licenses to 176 computer manufacturers and shut down many unlicensed manufacturers [China IT Market Report 1997].

6.1.3.1 The Business Computing Environment in China

The top five companies in market share for PCs in 1996 were Legend (9%), IBM (8%), Compaq (7%), AST (7%), and HP (6%) [China IT Market Report 1997]. In 1997, Legend remained the most popular brand in China, with a market-leading share of roughly 11 percent, ahead of second-place Compaq [Wallace 1998]. The price differential between brand-name computers and generic assembled PCs is enormous. A typical business computer, such as a Pentium 233-MHz MMX multimedia system with 32 megabytes of DRAM, a 15-inch monitor, a 2.5-gigabyte hard drive, a 24X CD-ROM drive, graphics, and fax/modem cards, would cost less than $800 in the Chinese assembly shops in Shanghai. Similar systems for Acer, NEC, or IBM would be 50 percent higher in price [Wallace 1998].

Most non-networked computers are dedicated to day-to-day operations and data processing, followed by financial management and multimedia applications. About half of commercial desktop PCs were non-networked in 1996; about 20 percent each were connected with NT and Novell, and the rest with WAN (7 %), Unix (3 %), and others. Networking is more common than system integration, but China's investment in system integration has been increasing at a phenomenal rate, spurred by the construction of the large government-funded "Golden" projects. The financial system comprises the biggest share of the system integration market, followed by commercial automation of store, stock, and cash register management, and then by intelligent buildings [China IT Market Report 1997].

CAD/CAM, process control, MIS/OA, and CIMS systems are becoming increasingly common and a focus of government investment in large and mid-sized "backbone" enterprises, textile design, and non-ferrous metal design applications throughout China. Other demands for specialized systems come from social service, education, publication, public security, and information processing institutions of various sizes.

China's leading domestic system integration enterprises are Legend, Nantian, Founder, Taiji, and Huasheng. Legend's 1995 system integration sales were $50 million, which rose to $80 million in 1996 from contracts in the fields of banking, taxation, transportation, energy resources, commerce, posts and telecommunications, environmental protection, broadcasting, and television; Nantian's sales exceeded $175 million in 1996 [China IT Market Report 1997]. Several of China's research institutes and universities have a sizeable share of the systems integration market, and twenty to thirty foreign-funded enterprises are also growing revenues from this field.   Foreign companies in this market include 3Com, IBM, Novell, SUN, Cisco, Intel, and Microsoft.

Problems in China's systems integration sector include insufficient transparency and fairness in the tender of large projects; insufficient capital, technical sophistication, and knowledge by management; and lack of regulatory laws, standards, and specifications. The government is trying to speed up the formulation of national and industrial standards and specifications for system integration and information servicing; to set up experimental enterprises and integration technology research laboratories, and to create a more transparent and fair business climate.

Other computer areas with growing importance in China's marketplace include the following:

- Mass market video conferencing;
- LAN management and security;
- Peripherals;
- IT standardization.

6.1.3.2   China's Home PC Market

The enhanced capabilities, declining prices, and market focus on PCs, combined with rising incomes in China, has made computers attractive to the home user. Almost non-existent before 1992, China's home PC market in 1996 was approximately 7.1 percent of total PC sales and growing rapidly, according to the New Century Group, a market research house [*Semiconductor Business News*, 25 June 1997]. The annual growth rate exceeded 50 percent from 1995 to 1997, and as of 1997, over 60 percent of all home PC owners had purchased their computers in the past two years [Carroll 1998b]. The number of PCs sold in the home market was about 340,000 in 1996 and about 540,000 in 1997 [China IT Market Report 1997].

The typical Chinese PC buyer is an urban resident earning over 40,000 yuan a year (~$4,813), who is buying the computer for educational use by a teenaged son or daughter. The New Century Group estimates that 60 percent of PCs purchased by home users are primarily for children's education, 30 percent for entertainment, and 10 percent for work at home. Although home purchases of computers generally represent a much larger investment for Chinese families than for families in more developed countries, the typical PC purchased for home use in 1996-7 was comparable to those purchased in developed countries: name brand, 486 or 586, over 16 MB RAM, 1 GB or more of hard disk space, multimedia-capable, with CD-ROMs and 28.8 or 36.6 kbps modems. Purchasers of Founder's home PCs are given Internet addresses and eight hours online free. In terms of brands, the general trend is that the large share of no-name domestic brands and store-assembled or clone PCs bought by Chinese consumers is declining, as is the share of foreign name brands, while the share of domestic name brands like Legend, Great Wall, Tontru, and Founder is increasing,

China's cultural emphasis on education as a means to social betterment, the one-child norm, the prestige of home computer ownership, and rising personal incomes are expected to continue to drive home PC sales at an annual growth rate of about 50 percent for several years. Despite this trend, a lack of quality software in the Chinese language, as well as cost and lack of consumer knowledge, still inhibit demand. Vendors are being pressed to reduce prices, improve software, and enhance support services, especially in remote areas. Influenced by these and other factors, the Chinese computer market has begun to place more emphasis on the value of software and services, as well as on hardware development and manufacturing.

As an example of the new emphasis on services, Legend started the "Legend Express" service system in December 1996, and Tontru started a similar "Green Service" in early 1997; these plans include such services as door-to-door servicing in selected big cities, 24-hour telephone consultation lines, one-time training classes for users (up to twenty hours), better-trained service teams, and other variations on technical support and equipment maintenance agreements. Another Chinese vendor, Lanchao Group, launched a "Service to Ten Thousand Homes" campaign in March 1997 that calls for building eight regional maintenance centers that will handle sales as well as repairs and after-sales service. It will offer three years' free service, one to three-day component replacement, and a three-tiered rapid maintenance plan. Such service plans are designed to help capture future as well as present market share by building brand awareness and loyalty [*China Telecommunications Weekly*, 24 March 1997].

6.1.3.3 Notebook Computers

There is much enthusiasm in the notebook computer market among a number of computer manufacturers in China. Firms with products in this market include Toshiba, MAX, IBM, Lunfei, AST, Compaq, Acer, NEC, Hewlett-Packard, Dell, DEC, Fujitsu, several Korean firms, and the domestic PC manufacturers Legend, Great Wall, Founder, and Tontru. Some 214,300 notebook computers were sold in China in 1997 — only about 4 to 5 percent of the sales of desktop computers (compared to about 40 percent of desktop sales in most developed countries) [Li 1998]. Notebook use in China is low. The higher price of notebook computers is a significant inhibiting factor [Roberts and Burbank 1998]. Costs, however, are falling somewhat due to the lower costs of LCD flat panel displays, and promotion and service are improving, so there is some optimism that this segment will have a growing market share. P/MMX-based notebook computers like P266 MMX, P200 MMX, and P233 MMX dominate at this time. They consume less power than Intel's new Pentium IIs.

6.1.4 Major Domestic Computer Manufacturers

China's major domestic PC manufacturers are Legend Holdings Ltd. of Beijing, Great Wall Group, Tontru Information Industrial Group, Founder Group, and Stone Group.

6.1.4.1 Legend Holdings Ltd. of Beijing

The Legend Group was established in 1984 by eleven research staffers from the Institute of Computing Technology, a branch of the Chinese Academy of Science [Zhang and Wang 1995]. Legend has become China's leading domestic information technology vendor since 1996. Its Legend Computer Systems Ltd., founded in March 1994, possesses large-scale manufacturing and markets three to four hundred thousand PCs per year [Legend Computer Systems 1998]. The Legend North System Integration Company is building the Nationwide Agriculture Information Network for the China Academy of Agricultural Science [*China Computer Trends*, 24 February 1997].

In 1993, Legend became the first Chinese PC maker to open a design center in California's Silicon Valley [Roberts and Einhorn 1997]. Taking a low-price strategy, Legend took the top spot in China's domestic PC market in the first quarter of 1997, in terms of total units shipped [*Semiconductor Business News*, 1 July 1997]. Legend is emphasizing an expansion of its domestic market share and will gradually shift operations to the international market. As of 1997, Legend has become an important supplier of PC motherboards worldwide. It exported motherboards to over forty countries, and its motherboard sales put the company in fifth place in the world in this market. Legend's total worldwide sales in 1997 reached a record $1.5 billion, and it expected sales to reach $2.25 billion in 1998. By the year 2000, Legend's ambitious goal is to raise its annual sales to $3 billion and to be among the world's top fifty computer firms, said the group's president, Liu Chuanzhi [Carroll 1998b].

In 1994, Legend had two thousand employees and thirty-four subsidiaries in China and overseas. Legend has set up over twenty-six branches around China with about a thousand sales and service agencies. This extensive distribution network has given it an important competitive advantage. Legend's extensive distribution networks in China have been effective in positioning its PCs. Legend's distribution arm started by selling many PC brands. That distribution group now pushed its own product lines in the market [Carroll 1998b]. In addition, Legend is using Hong Kong as the stepping stone to the Asia-Pacific market. Legend will use the build-to-order model to attract dealers to represent their products. Under the BTO model, dealers do not have to hold stock. As a result, dealers have lower operational costs and higher profit margins. Legend is expected to set up more than thirty dealers in Hong Kong to sell PCs to the government, education, and small and medium enterprises (SMEs) [Fu 1998].

Legend has consistently formed partnerships with foreign companies to acquire advanced technologies. For example, in March 1997, Legend Group became the first domestic PC vendor to sign a software OEM deal with Microsoft to distribute the simplified Chinese-language version of Windows95 [*China Computer Trends*, 24 March 1997]. MEI officials called the agreement the first deal to prove China is serious about intellectual property rights.

6.1.4.2   The China Great Wall Group

The Great Wall Group is a large state-owned enterprise with several significant subdivisions and joint ventures with IBM. One is the International Information Product Company, Ltd. (IIPC), founded in February 1994; another is GKI Electronic Product Company, Ltd., in operation since September 1995 [Dexter 1996]. In addition, Great Wall signed a license deal with Intel in May 1996 to manufacture Pentium motherboards. Great Wall was also the first company in China to create a server series based on Intel's Pentium Pro processor.

IIPC is one of IBM's six major PC production bases worldwide. It has five production lines, three offering IBM PCs labeled "Made in China," and two producing the Gold Great Wall series of PCs for domestic consumption [Li 1996]. The number of IIPC PCs under the brand names of Gold Great Wall and IBM reached one hundred thousand in 1995. The second joint venture, GKI, was established, based on the cooperative experience on IIPC. GKI was equipped with the world's most advanced overall plane welding technology. The enterprise has an annual production capacity of two million boards [Li 1996]. GKI thus became one of IBM PCs' major OEM suppliers.

Great Wall Group is not only producing IBM brand name products, but is also developing its own brands. In 1995, Great Wall was able to develop large-scale production of computers and a sales organization for monitors, terminals, disks, software drivers, video disk drivers, power disconnect switches, envelopes, and board cards [Li 1996]. In 1997 Great Wall introduced an ultra-thin notebook configured with a Pentium 100 and a multimedia notebook PC, both in the low $3000 price range [*China Telecommunications Weekly*, 24 March 1997]. In addition to producing parts and components for Great Wall PCs, the Great Wall group also supplies domestic markets with monitors, software drivers, and battery backup systems, and has become one of China's largest original equipment manufacturers (OEM) for do-it-yourself computers.

6.1.4.3   Tontru Information Industrial Group

The Nanjing-based Tontru Information Industrial Group is part of the key Information Industrial Group for China's Ministry of Electronics Industry. Tontru Group manufactures and markets PC-related products. By the end of 1996, Tontru had a full line of PC series available on the market – a business PC (Tongshi series), home PC (Tongle series), education PC (Tongxue series), PC server (Tongfei series) and portable PC (Tonguin series). Tontru has also produced Intel-based MMX PCs. As of 1998, Tontru had six production lines in Guangdong and Nanjing that manufacture

Pentium-type PCs, most using Intel chips (including the MMX chip), but some using 5X86 Cyrix chips. Tontru has twenty-eight offices and authorized maintenance centers. Half of its sales are in China's north and eastern regions [*China Computer Trends*, 3 March 1997].

In 1997, Tontru had become the second largest PC vendor in China behind Legend, with a selling capacity of more than two hundred thousand sets in the domestic market. Intel is Tontru's most important partner in China. In addition, Tontru Group has built close cooperation with Digital, Sunsoft, Samsung, LG, Leo, and Daewoo. The group has established six joint ventures and twelve branches, and undertakes R&D and production of monitors, keyboards, cards, power supplies, and cases.

6.1.4.4 *Others*

The Founder Electronics Group is controlled by Beijing University. Founder recruits top graduates for its three hundred-person research institute, which expects to add a hundred new employees annually during the Ninth FYP. Leader in Chinese software, Founder Group diversified into PCs in late 1995. In 1993 Founder was not among the top twenty PC vendors, but it zoomed to eighth in 1997. An estimated one hundred thousand units of PCs were expected to be produced in 1997 [Roberts and Einhorn 1997].

The Stone Group was established in the mid-1980s. It has ongoing cooperative relationships with Compaq and Mitsubishi Corporations of Japan. In 1996, the Stone Group was classified by *Business Week* as one of China's PC champions [Roberts and Einhorn 1997].

6.1.5 Major Foreign Competitors

China's PC market will remain competitive and challenging for both foreign and domestic suppliers. In 1998, major foreign computer manufacturers operating in China include IBM, AST, Compaq, Hewlett Packard, Dell, Gateway 2000, Apple, Sun Microsystems, Texas Instruments, and Packard Bell of the United States; NEC, Fujitsu, Hitachi, Casio, Oki, and Toshiba of Japan.

IBM's business in China dates back to 1934, when the company first installed a business machine for the Peking Union Hospital. IBM resumed its business in China in 1979 after the introduction of economic reform by the Chinese government. In 1992, it set up a wholly-owned subsidiary in Beijing, IBM China, to manage and coordinate all of IBM's marketing and production activities in China. IBM has set up seven joint venture companies in China since 1994. There are three manufacturing joint ventures: IIPC in Beijing (makes IBM and Great Wall brand PCs), Tianjin Advanced Information Products Corporation in Tianjin (manufactures banking peripherals and point-of-sale terminals), and GKI Electronics in Shenzhen (makes electronics cards and boards) [IBM China 1998].

In addition, there are two joint venture companies in software and system development: International Software Development Co. Ltd. (ISDC) in Shenzhen, and Advanced System Development Co. Ltd. (ASDC) in Beijing.

Two other joint ventures with domestic companies are in information technology services: The Blue Express Technical Service Company Ltd. (BE), and the Xun Tong Information Networking Research Development Company. The domestic partners of these IBM joint ventures include companies set up by Tsinghua University, Shenzhen University, the Ministry of Railroads, and three of the most important companies under the China Ministry of the Electronics Industry: The China Great Wall Computer Group, Kaifa Electronics Company, and JiTong Company [IBM China 1998].

Compaq was the fastest growing PC supplier in China. In 1993 it delivered eighty thousand PCs to the Chinese market. Compaq was the one of the leading PC vendors in 1996, with 7 percent of market share [China IT Market Report 1997]. AST is a major collaborator with the Legend Group. In 1993, it sold 140,000 desktop PCs in China, a market share of 30%. However, it only sold 145,000 PCs in 1996, which accounted for a market share of only 6.9 percent [China IT Market Report 1997]. AST entered the Chinese market in 1985 and set up a factory in Tianjin that reportedly produces more than one hundred thousand PCs annually [Zhang and Wang 1995]. AST became the number-one PC supplier in 1995, but due to severe competition from both domestic OEM PC makers and new foreign entries equipped with abundant resources and marketing skills, AST's significance in China's PC market has been reduced in recent years.

Dell Computer is the latest computer company to announce production plants in China to tap the booming market. By building production facilities close to customers, Dell hopes to boost its current one percent share of the Chinese PC market. The production is expected to launch in early 1999 [Robertson 1997].

6.1.6 PC Distribution Channels in China

China has been affected little by the Asian financial crisis and stands as one of the few growth markets in the region for PCs. The main beneficiaries of the growth in the PC market are the domestically produced models favored by local distributors, rather than foreign manufacturers [Carroll 1998b]. Local computer makers, particularly Legend, have retained the lead in market share, while U.S. and other foreign OEMs in the region, such as Compaq and IBM, have seen their presence slide or remain stable at best.

The critical issue is the implementation of a successful distribution channel strategy, especially in a place lacking well-established existing channels and trade transparency. As a result, foreign entries still rely on local distributors and dealers to sell their products, even if they have set up their own representative offices locally. Establishing joint ventures or strategic alliances with local manufacturers or even competitors is one way to distribute products successfully in China. For example, Legend manufactures its own PCs and also is the largest distributor for AST. Stone Computer Group produces its own brand-name PCs and also represents Compaq in China [*China Computer Trends*, 3 February 1997]. Dell may have an edge in terms of manufacturing, but its main challenges in Asia are related to channels. Dell's direct sales approach can only sell effectively to companies that have internal data processing capability. Dell is trying to build its reseller network, but faces difficulty working with the third-party channels [*Viewsletter IT Weekly* 1998].

The three basic ways in which computer systems are being purchased in China are face-to-face meetings or discussions, customer walk-ins, and fax/phone communications. The fax/phone method is not popular in China because customers and the communications infrastructure are generally less technologically sophisticated, and credit card ownership and acceptance is still uncommon.

The importance of channel partners is critical to success in doing business in a foreign country. A strong capability to provide local services and support is essential. Being able to establish a network of connections through local partners is even more crucial in China. The buying cycle tends to be lengthy in China, so a sound financial background and the ability to finance large purchases is also important. Overall, China's computer distribution channel infrastructure is different from those used in the industrialized nations. Face-to-face selling and buying methods continue to be the most comfortable method for many Chinese buyers [*China Computer Trends*, 3 March 1997].

6.2 SOFTWARE INDUSTRY

China's software industry is developing fast and has achieved a certain scale. However, the technological infrastructure and the commercial application market of China's software industry are still in the early stages of development. Having a large domestic market and a pool of engineers and scientists, the Chinese government is determined to transform its software industry into a global powerhouse.

6.2.1 Goals for China's Software Industry
China's goals for the Ninth Five-year National Development Plan and beyond include the following areas:

- expanding the market value of the software and information service industries to $6.5 billion by the year 2000 and increasing the number of Chinese software products to over 50,000; increasing the domestic market share of Chinese software products to 40 percent (33% in applications and 7% in system software) [CEIC 1997];
- promoting the industrialization of the software development process and popularizing the use of application software;
- increasing the percentage of large enterprises (e.g., retail sales, wholesalers, storage and warehousing, restaurants, industries) that fully utilize the computerized management system to 80 percent; the target

for the small and medium-sized enterprise (SMEs) is between 30 and 40 percent [China IT Market Report 1997];

- continuing the construction of a computer information service network (e.g., the Golden series projects, the financial information system, the national weather forecast system, and the national finance and taxation system) to cover the entire nation;
- developing an open Chinese-language platform on which Chinese software developers can design applications; the main purposes of this open platform are to
    - commercialize copyrightable application software to meet China's cultural, financial, and other needs;
    - develop China's information service industry;
    - extend the works of system integration, such as the Golden series projects;
    - expand scale of the software industry and encourage international cooperation.
- enacting and enforcing laws and regulations to protect intellectual property rights;
- Speeding up development and production of peripheral products and accessories, such as monitors, printers, IC cards, input devices, commercial ATMs, and data communication and networking products;
- to achieving international state-of-the-art technology levels in the information service industry by the year 2010 and becoming an international player;
- devoting major efforts to develop the domestic software industry, including setting up a number of software research institutes and development centers.

The Ministry of the Electronics Industry (MEI)[1] is the government body with the ultimate responsibility for China's software industry, although the State Science and Technology Commission also has some input. The two bodies jointly sponsor the China Software Industry Association (CSIA), which helps to regulate and promote the industry both at home and abroad. Other government organizations with an active role in developing China's software industry include the Institute of Software of the Chinese Academy of Science (ISCAS), and the China National Computer Software and Technology Service Corporation (CS&S).

---

[1] Note that MEI is one Chinese ministry slated to undergo major downsizing or consolidation with other ministries—in this case, merger with the Ministry of Posts and Telecommunications (MPT) and segments of the Ministry of Film, Radio and Television and of the Ministry of Aerospace and Aeronautics (those responsible for network system management) into the new Ministry of the Information Industry (MII) [Carroll 1998c].

6.2.2 Brief History and Overview of Software Development in China

The development of China's software industry started along with the development of the hardware industry in the late 1950s. The rate of growth has sped up since the economic reform began in 1979. The legislation and enforcement of the protection of intellectual property rights in 1991 has also contributed to the rapid development of the Chinese software industry.

Prior to 1979, the emphasis was on basic research rather than on product development, and no Chinese enterprise was producing and selling software. During the 1980s, China began to import foreign software while simultaneously developing and marketing indigenous products. Piracy of foreign software was prevalent. Before 1991 there was no copyright protection, and software piracy grew unrestrained, even in the export sector. In 1991, the Chinese government enacted two major regulations to protect computer software and other intellectual products. It demonstrated the willingness of the Chinese government to crack down on software piracy and to comply with international laws. The new laws did not entirely stop software piracy in China; nevertheless, the domestic share of software production increased significantly, and the rate of growth outpaced the growth rate of the national economy. The recent rapid growth of China's software industry has been stimulated by several national projects, such as the Golden series projects, that aimed at building national technological infrastructure and promoting the use of information technology.

Although China is interested in developing proprietary software of all types, it is also determined not to "reinvent the wheel" or to attempt to compete in saturated markets. China's main strength is in the development of Chinese-language application software for use in financial, banking, and industrial areas. China could also utilize its strength in developing Chinese-language information processing systems (including minority languages), publishing systems, and machine translation programs.

China's global competitive advantages in the software business are its huge market potential and quality work force. China has a large and growing number of qualified, relatively low-wage and hard-working software developers who could combine logical thinking with rich artistic imagination. China's economy and per capita income have been growing more rapidly than those of most developing countries since the 1990s, and are expected to continue this pace. In addition, government spending on building the national information infrastructure and the increasing number of people owning personal computers indicate a large potential market waiting for software manufacturers to explore. The challenges China faces in the coming years are [Yang 1997] to

- develop superior quality, unique products that are copyrightable and highly functional and responsive to the Chinese software environment (including software development tools);

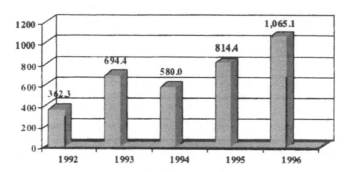

Figure 6.3 The Growth of China's Software Output Value ($million)
Source: China IT Market Report 1997.

- support Chinese information processing systems by adapting foreign software and creating indigenous software capable of adapting different platforms;
- take advantage of the changing technology and international product trends to make a unique contribution in areas such as multimedia and networking software.

At present, China's software industry is considered to be a "young" industry. In 1996, the number of software developers, technicians, engineers, and systems analysts was about five hundred thousand. The number is growing by about thirty-three thousand per year, due to students graduating from the universities [Yang 1997]. In terms of software education, China focuses on (1) training computer specialists; (2) providing non-computer specialists with comprehensive training in and familiarization with computers; and (3) expanding facilities to provide computer-aided education.

From 1985-95, China had an almost fourfold increase in the number of students pursuing various computer specializations. In 1995 the country had more than five hundred universities and colleges offering specialized computer studies; these graduated over ten thousand undergraduates, twelve hundred Masters specialists, and eighty-three Ph.D.s [Yang 1997].

China's software industry will continue to grow at high speed. Figure 6.3 shows the growth of the Chinese software market since 1992. The fluctuation between 1992 and 1994 is due to exchange rate conversion. The total output value in the software industry reached $1 billion in 1996. By the year 2002, annual sales are forecast to be at least $10 billion, according to Duh Jiabin, the general manager of Microsoft Corporation in China [Dexter 1996]. Local companies are expected to hold more than 50 percent of the Chinese market.

6.2.3 Intellectual Property Issues

The Chinese software market is still struggling with the issue of software piracy, a problem not only for foreign companies but also for Chinese firms. For small local companies, piracy can mean the end of business. For example, Beijing Kelihua, one of the top emerging software companies in China, suffered a huge loss during the sales of its first-generation educational software because pirates flooded the market with clone products. Small local companies are not alone. Foreign multinationals also face this problem. For example, it was possible to buy $12,000 worth of pirated U.S. software on CD-ROM for just $10 in 1996. The pirated versions of Windows95 were available at the same time the authentic version was realized in the United States.

Before economic reform began in the early 1980s, individual property rights, as well as intellectual property rights, were not respected and protected. Government regulations published in 1982 still claimed that all inventions and innovations belonged to the state [Zhang and Wang 1995]. However, the lack of attention to intellectual property rights is beginning to change. China's first copyright law, the "Copyright Law of the People's Republic of China," was put into effect in June 1991. In addition, "Regulations for the Protection of Computer Software Products" were also enacted in December 1991. China also acceded to the Berne Convention on October 15, 1991 and the World Copyright Convention on October 30, 1991 [China IT Market Report 1997]. These movements gave a strong push to the development of China's software industry.

6.2.4 Major Software Technology Development Institutions in China

The Chinese government has funded several research organizations to promote software technology development and applications. This section introduces two major software research and development institutes under the umbrella of the Chinese Academy of Sciences, and one state-owned software technology company.

6.2.4.1 Institute of Software

Established in 1985, the Institute of Software, Chinese Academy of Science (ISCAS), is a comprehensive academic institution with its major focus on researching the fundamental theory of computer science and the development of advanced technology software. Major research areas include computer science theory, software engineering, parallel computing, multimedia, and mathematical logic. ISCAS also conducts software product development in the areas of Chinese language processing, system integration, compiling and testing, as well as business applications software. ISCAS has established extensive long-term cooperation with the international science and technology community and industries. Cooperative research joint ventures have been set up with companies such as IBM/Lotus, NEC-CAS Software Lab., and Microsoft [ISCAS 1998].

6.2.4.2 The National Research Center for an Intelligent Computing System

Founded in March 1990, the National Research Center for an Intelligent Computing System (NCIC) is a research and development center for advanced computer technology under the Chinese national high-tech 863 program. NCIC is supervised by the State Science and Technology Commission, and administratively supported by the Institute of Computing Technology (ICT). The main objectives of NCIC are (1) to design and implement marketable symmetric multi-processors (SMP) and massively parallel processors (MPP); (2) to develop innovative products by integrating research results generated from the 863 program; (3) to perform fundamental research on high-performance computers and intelligent computing systems; and (4) to carry out international cooperation and academic exchange [NCIC 1998].

6.2.4.3 China National Computer Software and Technology Service Corporation

The China National Computer Software and Technology Service Corporation (CS&S) is a science, industry, and trade integrated high-tech enterprise specializing in commercial software development, system integration, and computer application engineering and technology services. It is a state-owned enterprise under the control of MEI, and is composed of many subsidiaries. CS&S has fixed capital of more than $25 million. It has established various kinds of cooperative relationships with many multinational corporations, such as IBM, Novell, Microsoft, NEC, and Complex of Singapore [CS&S 1998].

6.2.5 Construction of Software Parks

The Chinese government has constructed a number of software development parks and software production bases. As with specialized economic and technology development zones, preferential policies have been adopted to assist software startups located within the park. In addition to the Northern Software Base in Beijing, the other two software parks established since 1992 are the Shanghai Pudong Software Park and the Southern Software Industrial Park in Guangdong Province. Furthermore, the China Computer Software National Engineering Research Center has been founded in Northeastern University [China IT Market Report 1997]. The building of software parks and other software research and development centers will surely contribute to the development of China's software industry.

6.2.5.1 Shanghai Pudong Software Park

Shanghai Pudong Software Park (SPSP) Development Co. is a new enterprise established jointly by the China Ministry of Electronics Industry (MEI) and the Municipal Government of Shanghai. SPSP was established in July 1992. It takes up 300,000 square meters of land space and will accommodate ten thousand software engineers upon completion. Software-related projects open for investment and cooperation include information processing services, software system integration, data communication and network technologies, and other software-oriented businesses [SPSP Development Co. 1998].

6.2.5.2 Southern Software Industrial Park

The Southern Software Industrial Park (SSIP) is jointly established by the MEI and Zhuhai Municipal Government. Phase 1 of the project has been completed, and phase 2 will be executed between 1998 and 2000. Phase 1 provides a capacity for fifteen enterprises engaged in the development of software. The annual turnover for software is estimated to be $50 million. Enterprises that engage in the development, production, sales and services of software, hardware, system integration and relevant products can apply for admission [SSIP 1997].

6.2.6 Major Problems in Chinese Software Industry

Major challenges facing the future development of the Chinese software industry can be summarized as follows:

- software production is not yet industrialized.
- the software market system is not taking shape and software products are not highly commercialized.
- the technological infrastructure for software development is inadequate.
- the state's investment in the software industry is not sufficient.
- the knowledge of software users (both industrial and end users) is insufficient.
- intellectual property rights are not respected.

Fortunately, the Chinese government has acknowledged and addressed these issues through the implementation of numerous policy instruments to build a proper industrial infrastructure for software development and applications. However, more public and private efforts are required to advance the software market in China.

6.3 KEY RESEARCH PROJECTS FOR THE FUTURE

Several key information technologies have been identified and targeted by the government agencies and research institutions to be developed to modernize the technological infrastructure of China's information electronics industry.

6.3.1 Computer Technology

The key computer technologies selected by Chinese authorities to be improved and developed within and beyond China's Ninth Five-year National Development Plan are
- technology for new types of processors;
- high-performance computing technology (SMP and MPP);
- computer networking technology (network system architecture, data transmission and exchange, network addressing protocol, network interconnection, network management, and network security);

- mobile computing technology (structure of mobile computer system, interface standards and execution technology, wireless communication, LAN/WAN interface, pen input and voice recognition technology);
- multimedia information processing technology (multimedia data compression, searching, recovery and transmission technology);
- system security technology (system security and execution methods, security analysis, appraisal and test technology, information divulge-prevention-technology, a mechanism-of-evil program and its confrontation technology);
- computer peripheral equipment and technology (optical storage technology, flat panel display technology); and
- new application of remote sensing and computer network interconnection technology;

6.3.2 Software Technology
The key software technologies to be developed are
- open system and platform technology (transplantation and mutual manipulation);
- CASE (computer-aid software engineering) tool technology (structural analysis of system demand, program coding, online picture intersection, coding automation, automatical source code generating);
- LOOP technology (block multiplex and patching technology, microcore operating system);
- industrialization of the software production process (structure of software system, developing method, quality assurance system, measuring and appraisal method);
- exploitation techniques for system software (operating system, language processing system, and database management system);
- software techniques for inputting and processing Chinese characters (Chinese processing and software platform);
- multimedia software technology (multimedia operating system, multimedia database, multimedia developing tools and system).

In sum, the program for developing software technology will emphasize applications and service software. The point of developing and using application software is to increase productivity, enhance managerial capability, and improve manufacturing and delivery of products and service to customers. The major emphasis in the development of service software is on software integration.

Chapter 7

CHINA'S TELECOMMUNICATIONS INDUSTRY

China's telecommunications industry has made significant progress over the past two decades. The national telephone density increased from 0.43 percent in 1980 to 8.11 percent in 1997 [Liang, Zhang, and Yang 1998]. The number of phone lines per hundred people was 1.5 in 1993, compared to Hong Kong's forty-eight and Taiwan's thirty-seven. By the year 2000, the number of phone lines per hundred in China is expected to reach five [Ure 1995]. The public telecommunications network has developed faster than that of any other country in the mid-1990s, thanks in large part to increased competition in the domestic market [Rehak and Wang 1996]. Competition from both foreign vendors and domestic groups is helping to spur modernization of China's telecommunications industry. This chapter begins with a brief introduction of China's telecommunications industry, then discusses the development of China's data communications network, the increasing popularity of the Internet, and the growth of the mobile communications market. Major issues regarding Chinese-foreign joint ventures and the prospects of China's telecommunications industry are addressed in the final two sections.

7.1 THE DEVELOPMENT OF CHINA'S TELECOMMUNICATIONS INDUSTRY

China has rapidly developed an advanced telecommunications structure, driven by economic reform, global trade growth, and modest liberalization. Since the implementation of the open door policy in the early 1980s, the size of the telecommunications network has increased steadily, and the level of technological sophistication and communications capacity have expanded considerably.

7.1.1 Periods of National Telecommunications Development
China's telecommunication industry has undergone three major periods of development since economic reform and the open door policy began in 1978 [Liang, Zhang, and Yang 1998].

7.1.1.1   The First Period – 1978 to 1984

China's economic reform, aimed at transforming a command-and-control planned economy into a so-called "market socialism," meant a need for rational and coordinated production and supply systems. In addition, the open door policy that encouraged direct foreign investment in many areas of the country required a more developed communications system. In order to speed up the pace of reform and modernization, the Ministry of Posts and Telecommunications (MPT) set the first priority as modernizing China's telecommunications infrastructure. During this period, MPT unified its management control system, increased the telephone capacity in large cities, lowered telephone installation fees for the first-time user, and actively introduced foreign funds, advanced telecommunication technologies, and modern management methods to China.

7.1.1.2   The Second Period 1985 to 1989

As China made a commitment to economic reform and opening domestic markets for foreign investment, telecommunications service has become ever more significant for the development of the national economy. During this period, telecommunications was recognized as a strategic economic pivot, aggressively promoted by the central government through the implementation of several preferential policies approved by the State Council and supported by local governments. For example, the "three 90 percents" strategy was inaugurated in 1988 to benefit the MPT's efforts to develop the telecommunications infrastructure. This policy stated that (1) 90 percent of central government loans for telecommunications did not have to be repaid; (2) provincial telecommunications authorities could keep 90 percent of their taxable profits; and (3) the MPT could keep 90 percent of its foreign currency earnings from international traffic [Ure 1995][1]. These policies had accelerated the construction and transformation of China's telecommunications network.

7.1.1.3   The Third Period 1990 to present

This period has been marked by rapid achievements in China's telecommunication industry. Table 7.1 shows the tremendous growth of telecommunications service and capacity since 1990. In 1997, there were over seventy million telephone subscribers in China, and more than fifteen million new telephone subscribers were able to access the public switched telephone network (PSTN). The switching capacity of the central office exchanges reached more than 110 million main lines in 1997. According to the newly established Ministry of the Information Industry (MII), China's telephone network capacity has reached 121 million main lines in the second half of 1998 [*Beijing Review*, 26 October 1998, p.26]. The number of new mobile phone and paging subscribers also increased significantly year by

---

[1] The policy was later repealed in 1995 because it was deemed no longer necessary [Harwit 1998].

year after 1990. By 1997, China's telecommunications network had developed into the world's second largest network in terms of overall capacity [Liang, Zhang, and Yang 1998, *Beijing Review*, 4 May 1998, p.21].

Figure 7.1 shows the growth in telecommunications investment versus overall GDP growth, a fivefold increase from 1990 to 1994. However, the rate of increase seems to have leveled off after 1994.

Another major change in this period was the structural reorganization of the industry toward decentralization. The central government began gradually ending central control and moving toward a system of open competition. Before 1990, the provincial offices took direct orders from the

Table 7.1
Development of Telecommunications Service and Capacity
(Unit: Thousand)

Telecommunications Service	1990	1995	1996	1997
Long distance telephone circuits	112	865	1,040	1,240
Automatic toll switching capacity	161	3,408	4,260	4,450
Central office exchanges	12,010	70,960	93,180	110,900
Telephone subscribers	6,850	40,710	54,950	70,270
Mobile subscribers	18	3,630	6,850	13,230
Paging subscribers	437	17,430	25,410	34,190

Source: Adapted from Liang, Zhang, and Yang 1998.

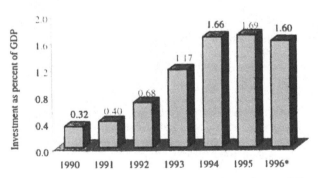

Figure 7.1 China's Fixed-asset Investment in Telecommunications as Percent of Gross Domestic Product (GDP), 1990-1996
* Estimated.
Source: Adapted from Harwit 1998

MPT; since 1990, the provincial and municipal Posts and Telecommunications Administrations (PTAs) have had autonomy in making planning and investment decisions, as well as accounting for their own performance, though MPT retains authority over network standards. The result of this change has added to the industry's efficiency [Harwit 1998].

7.1.2   The Telecommunications Market in China

As indicated in Table 7.1, China's telecommunications market has been developing rapidly since 1990. Total revenues were $2.5 billion (13.4 billion yuan) in 1991 and $10.5 billion (87.3 billion yuan) in 1995. By 1997, total revenue from telecommunication service reached $20 billion (163.4 billion yuan), an increase of more than 35 percent over 1996.

7.1.2.1 Telephone Service

The number of subscribers for traditional telephone service and mobile phones, as well as pagers, has increased enormously since 1990 (see Table 7.1). In 1998, the national and urban penetration rates densities had risen to 9.5 and 27.5 percent, respectively [*Beijing Review*, 26 October 1998, p.26]. The gap between rural and urban telephone subscribers has also begun to close. There were about 3.7 times as many phones in the urban areas as in rural areas. By 1996, the ratio of urban to rural telephone subscribers had dropped to 3.5 [Harwit 1998]. The increase in rural telephone subscribers has spurred the growth of township and village enterprises and MPT's preferential policies favor the less developed regions in China.

7.1.2.2 New Value-added Telecommunications Services

Many new value-added telecommunications services have been introduced to the Chinese market in recent years. Table 7.2 shows the development of these new value-added services since 1994. Except for videotext and electronic data interchange (EDI) services, all new services experienced triple-digit annual growth in 1997, especially the Internet service, which did not exist in 1994, but increased more than 377 percent over 1996.

7.1.3   Competition in China's Telecommunications Market

The Ministry of Posts and Telecommunications (MPT), under the guidance of the State Council, held a monopoly over all telecommunications services in China until 1994. The MPT had complete responsibility for both the regulation and the operation of networks and was responsible for policy, equipment approval, planning, and interprovincial network construction and operations. However in 1993, a new division, the Directorate General of Telecommunications (DGT), was spun off from the MPT to handle telecommunications network operation and maintenance [Rehak and Wang 1996]. The DGT was later separated from the MPT in early 1994 to become an independent enterprise. In March 1995, the DGT officially became China

Table 7.2
Development of New Value-added Telecommunications Services
(Unit: Thousand)

New value-added service subscribers	1994	1995	1996	1997	Annual growth rate (%)
Packet switching	8.5	28.0	56.4	84.6	115.1
DDN	-	17.0	51.4	11.7	156.3
E-mail	2.3	5.9	10.0	14.7	151.4
Videotext	1.1	2.5	2.0	2.3	27.5
EDI	-	0.13	0.1	0.2	27.9
Facsimile storage transfer	-	0.50	1.2	2.9	142.6
Internet	-	7.0	34.0	159.8	377.8
Frame relay	-	-	-	3.1	-
N-ISDN	-	-	-	0.3	-

Source: Adapted from Liang, Zhang, and Yang 1998.

Telecom when it registered with the State Administration of Industry and Trade as a legal entity. China Telecom is the largest telecommunications carrier in the country, responsible for operating and building a nationwide telecommunications network and providing basic and value-added services. By the end of September 1998, China Telecom boasted 104.4 million phone subscribers. Of those, more than 20.8 million were new subscribers in the first three quarters (63 percent desk phone and 37 percent mobile phone) [CEInet IT News, 25 November 1998].

The growth in China's telecommunications market has been particularly rapid since the formation in 1994 of China United Telecommunications Corporation (Unicom). Unicom was established to provide all kinds of telecommunications services, challenging the monopoly in China's telecommunications service market.

China Unicom is a consortium authorized by the Chinese government to compete with the MPT in all spheres of telecommunications operation. Formed by the Ministry of the Electronics Industry (MEI), the Ministry of Railways (MOR), and the Ministry of the Power Industry (MOPI), China Unicom also received financial support from thirteen other influential Chinese investors, including China International Trade and Investment Corporation [Rehak and Wang 1996]. The MEI had manufactured equipment for the telecommunications industry, and also wanted to join the operation, but lacked its own telecommunications lines. The major motive for the MOR and the MOPI to support the creation of China Unicom was that both operate extensive private telecom networks, and they hoped to utilize their excess communications capacity by offering public network services.

By mid-1995 Unicom had invested an estimated $82 million in network installation nationwide, with a plan to install a national trunk network based on the private network infrastructure of its shareholders [Rehak and Wang 1996]. The initial goals of China Unicom were to achieve 10 percent of both domestic and long-distance service and take 30 percent of the mobile communications domestic market by the year 2000 [Harwit 1998].

The competition between China Unicom and China Telecomm has started to change the market dynamics and sped up the growth of China's telecommunications service market. A new Ministry of Information Industry (MII) was established in March 1998 to replace the former MEI and the MPT, and to take over network management functions from the Ministry of Radio, Film, and Television. The MII's main function is to plan, regulate, and control the development of the information and telecommunications industry. However, the consolidation of the MEI and the MPT may cause a negative impact on competition because now the MII oversees both China Telecom and China Unicom. However the new MII charter spelled out by the National People's Congress in March 1998 included the dictate that the ministry should reasonably allocate resources to avoid duplicative construction and safeguard information security [Robertson 1998a], it remains to be seen how this merger between two former competitors will affect the level of competition in China's domestic telecommunications service market.

7.2   THE DEVELOPMENT OF CHINA'S DATA COMMUNICATIONS NETWORKS

Data communications networks, unlike the traditional tele-communications infrastructure, permit multimedia communication (voice, text, and graphics) between devices connected by telephone lines, microwaves, and satellites. As indicated in Table 7.2, the growth of data communications value-added services in e-mail, fax, and on-line information systems, such as the Internet, have increased exponentially since 1995. The market for data communications hardware, such as modems, routers, and servers, will also grow rapidly in the years to come. This provides an excellent opportunity for foreign manufacturers to supply China's networks with equipment and related software. This section discusses the construction of China's data communications networks and of Internet working units to provide Internet services in China.

7.2.1   Data Communications Networks

Established in 1994, the Data Communications Bureau (DCB) of the former Ministry of Posts and Telecommunications (MPT) is in charge of administering China's national public data communications networks. DCB's duties include planning, constructing, operating, and managing national data communications networks and related services. There are four major networks that are currently under the supervision of DCB: (1) China Public Packet Switching Data network (CHINAPAC); (2) China Public Digital Data network (CHINADDN); (3) China's main Internet backbone, CHINANET; and (4) DCB's frame-relay networks. DCB's frame-relay networks offer many benefits to users such as bandwidth-on-demand capabilities, error-free transmission, and lower management and infrastructure costs [Crisanti 1997]. As of July 1997, CHINAPAC had 140,000 installed data communications terminals and 70,000 users. CHINADDN had 190,000 terminals and 80,000 subscribers, and CHINANET had about 80,000 Internet users [Crisanti 1997].

In addition to national public data communications networks, a number of organizations have operated private networks for internal communications. For example, the Ministry of Power Industry, the Ministry of Communications, the Ministry of Railways, and the People's Liberation Army all have their own private communications systems, independent of DCB's public networks.

The competition in China's data communications market is similar to that in the telecommunications market. Jitong Communications Corporation was established to be a MPT competitor in networking services, just as Unicom was set up to compete directly with the MPT in the telecommunications market. In practice, the MPT retained a significant amount of control over both the voice and the data communications infrastructure in China. In order to level the playing field, the Leading Group on Information, a supra-ministerial body directly under the supervision of the State Council, was established in May 1996. The Leading Group's responsibilities include drafting policy to regulate China's information industry. However, it is unclear whether the consolidation of the MPT, the MEI, and several other ministries into the Ministry of Information Industry will hinder or increase the level of competition in China's data communications market.

Overall, China's data communications networks suffer from poor management, which causes inefficient and redundant use of equipment. In addition, limited competition gives rise to higher prices for end users [Crisanti 1997].

7.2.2   The Use of the Internet in China
The first network in China cooperating with the outside world was the Chinese Academy of Sciences Network (CASNET), which was established by the Network Unit center (NCFC) of the Chinese Academy of Sciences in 1988. CASNET is the Chinese Academy of Sciences' national research network.

The Chinese Education and Research Network (CERNET) is the first nationwide education and research computer network in China. The CERNET project is funded by the State Planning Commission and managed by the Chinese State Education Commission. The main objective of the CERNET project is to establish a nationwide network infrastructure to

Table 7.3.
Internetworking Units in China

Inter-networking Unit	Network Name	Number of Users	Number of Networked Hosts	Number of Networked Units	Band-width (kbps)
Tsinghua University	China Education and Research Network (CERNET)	50,000	3,000	110	128
China National Network Information Center	Chinese Academy of Sciences Network (CASNET)	15,000	114	120	64
Jitong Company	China Golden Bridge Network (CHINAGBN)	1,600	160	37	256
DCB of the Ministry of Posts & Tele-communications	CHINANET	64,866	N/A	20	256

Source: CEInet China IT Market Report 1997.

support education and research in and among universities, institutes, and schools.

In addition to the two academic research networks, there are two public commercial networks in China. CHINANET was started in 1994 by the MPT to provide various Internet services to public users and to promote the commercialization of the information network. CHINAGBN is a nationwide, public, economic information-processing network sponsored by the MEI. The network was first implemented in March 1993, and is one of the most important components of China's Golden Bridge Project. The major objectives of the CHINAGBN are to establish an economic information network to interconnect heterogeneous private networks of multiple departments and sectors, and to establish a computer information service system for government agencies and private enterprises.

There are local networks with some nodes connected to the Internet. Table 7.3 shows the structure of Internetworking in China. The four networking units mentioned above have been approved by China to provide Internet services.

In China, subscribers to the Internet have been increasing at a growth rate of more than 300 percent between 1995 and 1997 (see Table 7.2). Many companies are attracted to this potentially lucrative market. There are about a hundred Internet service providers (ISPs) providing electronic mail, document transmission, world wide web browsing, and remote login service. Those using the Internet are mainly government officials, technology buffs, academics, and business people. Most connect to the Internet through CHINANET, which is currently run by the newly established Ministry of Information Industry [Krochmal 1998a].

The State Information Office, China National Network Information Center (CNNIC), and the CNNIC Committee recently decided to release statistics every January and June regarding Internet access, such as the number of computers linked with the Internet, the number of users, their distribution, the information flow, and registered web-site names. According to the CNNIC, in the first half of 1998 China had 1.18 million Internet subscribers, almost twice as many as were registered at the end of 1997. The number of computers integrated with the Internet increased to 542,000 from 299,000 on October 31, 1997, of which the number directly linked rose from 49,000 to 82,000. There are 9,415 area names registered in ".cn" (CHINANET) and 3,700 web-sites: 72.6 percent for domestic regions, 11.8 percent for international, and 15.6 percent shared [*Beijing Review*, 2 November 1998].

Major existing problems with Internet usage in China include(1) too few homepages in Chinese; (2) of the several large, domestic networks except CERNET and CSTNET, which are connected to each other, most are not connected, resulting in inconvenient access; and (3) Chinese browsing platforms are incomplete, with poor Chinese translation accuracy [China IT Market Report 1997].

Chinese government leaders in Beijing see the Internet not only as a tool for economic development but also as a threatening source of politically sensitive information [Roberts 1996a]. China believes the Internet is a crucial force for opening its doors to the outside world, but Internet users in China will see content that has been selected, filtered, and published from government servers in ten Chinese cities [Krochmal 1998b]. With a relatively small number of accounts currently in service, censors can search for key words or web-sites considered dangerous by the government. But as the number of web-sites grows, the Internet will become more difficult for the government to monitor.

The Chinese government has set up several companies to manage the Internet. For example, the China Internet Corporation (China.com) is working with America Online (AOL) to build a co-branded web-based service, guided by terms of service conforming to Chinese laws. China Internet Corporation is also collaborating with AOL and Netscape on a Chinese browser that incorporates ratings to filter net content [Krochmal 1998b].

China is aggressively expanding its network infrastructure to support the widespread demand for Internet use in the next century. However, there will be a shortage of content for Chinese users. China is seeking to increase the content of Chinese-based web-sites by signing deals with Bloomberg, Bertelsman, Netscape, Yahoo, and AOL, among others. China.com selects the content it wants, translates it, and posts it on the web [Krochmal 1998b].

7.2.3   Emerging Electronic Commerce in China

The increasing popularity of on-line information systems in China will facilitate the growth of electronic commerce (e-commerce) in transactions between businesses and between businesses and end users. In May 1998, IBM Global Services and the Ministry of the Information Industry (MII) signed an agreement to promote e-commerce in China [Niccolai 1998, Haber 1998]. According to the agreement, IBM will work with China Telecom to provide customer billing and services to establish managed network services. IBM will also provide China Telecom with business and technical consulting expertise as it builds a new business transaction model [Haber 1998]. In addition, IBM will also work with Chinese companies and software developers to produce Chinese versions of JavaOS for Business and Lotus Notes. The agreement will help augment local Java programming skills, as well as ignite China's e-commerce software industry [Niccolai 1998].

IBM has begun providing e-commerce solutions in several of China's provinces through its nine branch offices, eight joint ventures, and one research facility. For example, IBM is in the initial stages of working with Hunan Telecom on a system that will eventually result in an inter-city e-commerce network [Haber 1998].

Another example of an e-commerce project under way at IBM Global Services in China is the development of the first electronic shopping center and payment system for Guangdong Telecom. In addition, IBM has also been providing the skills and technical expertise to China Telecom to build a next-generation customer care and billing system that will manage services, collections, and systems and networking management [Haber 1998].

The future growth of e-commerce in China looks promising as more and more people can access on-line information systems. Chinese businesses and consumers will gradually adopt the new business model to take advantage of the benefits of convenience and low transaction costs. However, the issue of on-line transaction security and the popularity of using credit cards and personal checks remain as obstacles to the implementation of e-commerce in China.

7.3   MOBILE COMMUNICATIONS

In the 1990s, China's mobile communications industry has developed rapidly. Mobile telecommunications and paging services are developing much faster than traditional telecommunications services in China. This section describes China's mobile communications market, networks, and competition, and its development strategy for the future of the industry.

7.3.1   Growth of China's Mobile Communications Market

Since the early 1990s, the number of mobile telecommunications subscribers has soared by 80 to 150 percent annually. Figure 7.2 shows the annual net growth of China's mobile phone subscribers since 1991. The

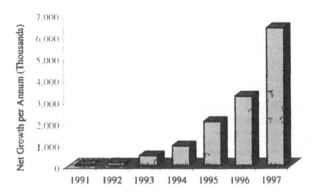

Figure 7.2 Net Growth per Annum of China's Mobile Phone Subscribers, 1991-1997
Source: CEInet China IT Market Report 1997

annual demand for the domestic mobile telecommunications market is about $7.3 billion (60 billion yuan), about two-thirds of which is for cellular mobile telephones. The total number of mobile phone subscribers is expected to exceed 30 million by 2000, with an average penetration rate close to 3 percent [Liu X. 1998].

The huge potential growth of China's mobile communications market has attracted major international manufacturers. As a result, competition has started to heat up among domestic and foreign equipment producers. As of 1998, China has more than 120 mobile telecommunications equipment manufacturers, many of them still using outdated technology and traditional operational mechanisms. Therefore, China must rely heavily on imports of mobile communications equipment to satisfy the increasing domestic demand for services.

7.3.2 Mobile Communications Networks in China

Stimulated by the huge demand for service, China began to build up its own manufacturing capabilities and network systems. The Ministry of the Information Industry (MII) is heavily involved in the effort to transfer foreign technologies, develop digital mobile communications equipment, and construct mobile communications networks. As of 1998, China has three major mobile telecommunications networks.

7.3.2.1 Analog Mobile Communications Network (TACS)

Total access communications systems (TACS) represent the industry standard for analog cellular telephone communications. TACS operates in the 900 MHz range. There are two 900 MHz TACS networks in China: Network A covers twenty-one provinces and cities with Motorola equipment; and Network B covers fifteen provinces and cities with equipment from

Sweden Ericsson. Nationwide roaming[2] of both Network A and Network B were completed in 1994. Interconnection and roaming between the two networks were realized in 1996 by China Telecom. As of 1997, China's analog communication network has covered 1,460 counties in thirty provinces and cities nationwide [China IT Market Report 1997].

7.3.2.2 Digital Mobile Communications Network (GSM)

The global system for mobile communications (GSM) is a pan-European standard for digital cellular communications. GSM is now emerging as the dominant digital standard in Asia. The completion of the GSM network in Lhasa (the provincial capital of Tibet) and its interconnection with the Chinese GSM network began nationwide interconnection. As of 1997, GSM nationwide roaming covered more than fourteen hundred Chinese counties, and international roaming with more than thirty countries has been realized.

China Unicom's GSM network is the second digital network in China covering more than a hundred cities with two million or more users [China IT Market Report 1997]. China Unicom launched GSM service in July 1995 with its own four operational GSM networks (at the same time, China Telecom has fifteen GSM networks covering the whole country). To generate revenues quickly, China Unicom is focusing first on cellular services and has been a driving force behind the spread of the GSM digital cellular standard in China. By 2000, China Unicom's GSM network will be able to cover the whole country.

7.3.2.3 CDMA Mobile Communication System

Code division multiple access (CDMA) is an American digital standard approved by the U.S. Telecommunications Industry Association in August 1993. The standard promises to deliver far greater capacity gains than GSM, and is expected to require fewer cell sites to achieve the same coverage. In order to ease the tension in frequency resources, China introduced a CDMA mobile communications system in 1997. China Telecom started building a CDMA experimental network in the four cities of Beijing, Shanghai, Guangzhou, and Xian. The CDMA experimental network has been implemented in Beijing and several provinces, including Fujian, Jiangsu, Shandong, Hubei, and three northeast provinces.

7.3.3 Competition in China's Mobile Communications Market

The high growth rate of mobile subscribers in 1997 could be attributed partly to the competition in the mobile service market between China Telecom and China Unicom, and partly to the decrease in tariffs on mobile communications service. Overall, China Telecom is still the leading operator in the market. Its market share was 97.4 percent in 1997. China Unicom is only operated in eighteen of the total thirty-one provinces in China [Liang,

---

[2] Roaming technology allows cellular phone subscribers to make and receive their calls outside of their designated service areas.

Zhang, and Yang 1998]. China Unicom has consistently complained that China Telecom has obstructed its business by delaying interconnection agreements. China Unicom was supposed to challenge China Telecom's market domain, but lack of crucial support from the State Planning Commission has hampered its progress. Worse, in 1997 several foreign companies have stopped cooperating with the company, citing a lack of returns.

A third mobile communications service provider, Great Wall Mobile Communications, was established in 1997 to compete against China Telecom. Great Wall has attracted much foreign investment from companies such as Motorola, Lucent Technologies, Samsung, and Nortel. Great Wall is also building a digital mobile communications network against the mostly analogue networks of China Telecom. The difference is that Great Wall is applying the CDMA digital standard promoted by the U.S., as against the European-supported GSM standard adopted by China Unicom. The Great Wall Network, China's third honeycomb network, is operating in Jiaxing, Fanyu, Beijing, and Xi'an on a trail basis. Great Wall has a higher chance of success than China Unicom because it is a fifty - fifty joint venture between the MPT and the People's Liberation Army. Great Wall plans to complete the deployment of ten to twenty CDMA networks in coastal cities in 1998 [*Asian Business Review*, December 1997, p. 52].

7.3.3.1   Foreign Suppliers of Mobile Communications Equipment

There were more than seventeen foreign vendors in China's mobile phone market in 1997, including Motorola (43 percent market share in 1996), Ericsson (34 percent), Nokia (8 percent), Siemens (7 percent), and NEC (5 percent) [China IT Market Report 1997]. China's analog cellular mobile telephone system equipment, such as the 900MHz TACS mobile switch, base station, and network control equipment, are mainly supplied by American Motorola (Network A) and Sweden Ericsson (Network B). Major suppliers for China's GSM system equipment are Ericsson, Nokia, Nortel, Alcatel, Siemens, Itatel, and Motorola. As for the advanced CDMA network system equipment, Lucent Technologies, Motorola, Nortel, and Samsung are the major vendors.

7.3.3.2   Domestic Manufacturing Capabilities

Between 1991 and 1995, the Chinese government offered the cities of Nanjing, Shanghai, Wuhan, Guangzhou, and Hangzhou financial and technological assistance to construct mobile telecommunications equipment manufacturing bases [Liu X. 1998]. In addition, many domestic enterprises have developed their own mobile telephone switching, transmitting, and connection systems. The development of domestic manufacturing capabilities in communications equipment and network systems is inevitable, given the growing demand for telecommunications services and Beijing's efforts to encourage local manufacturers. However, it may take at least one decade before the quality and reliability of Chinese-made equipment will threaten the foreign suppliers of leading-edge communications technologies and systems such as CDMA, asynchronous transfer mode (ATM), synchronous digital hierarchy (SDH), and videoconferencing.

7.3.4   Development Strategy for Mobile Communications in China

China's near-term targets for developing mobile communications networks include [*CEInet IT News*, 24 November 1998]:

- Small-area optimization distribution will be strengthened for TACS analogue networks already constructed, and base stations will be added to rationalize coverage.
- Post and telecommunications enterprises will try to increase the mobile communications coverage rate to more than 95 percent in all counties in eastern and central China, and over 30 percent in the western region.
- Telecom operators will increase efforts to develop GSM digital mobile communications system, expand the capacity and coverage of the GSM network, and improve the quality of transmission.
- A dual-band GSM900/1800 system will be set up to ease the intensified frequency supply in some areas, while the building of the pilot CDMA mobile telecommunication network will be continued.
- China will follow the development of the third-generation mobile telecommunications and IMT-2000 closely, and try to build the capability to design and manufacture its own IMT-2000 system as soon as possible.

7.4   FOREIGN JOINT VENTURES IN CHINA'S TELECOMMUNICATIONS MARKET

Every major international telecommunications company is pursuing opportunities in China, all hustling to capture a share of China's huge potential market. Until 1996, foreign telecommunications companies were allowed to provide only equipment and handsets, while China Telecom and China Unicom had a domestic monopoly over both fixed-line and cellular networks. In 1996 the Chinese government permitted seven foreign switching suppliers to manufacture locally and sell to MPT [Rehak and Wang 1996]. Since then, large transmission equipment manufacturers either have set up or are planning to set up joint ventures to manufacture SDH fiber-optic transmissions systems in China.

China National Posts and Telecommunications Industry Corporation (PTIC), a state-owned enterprise, is promoting joint ventures among local and foreign manufacturers of equipment to produce such items as switches, optical communications gear, and mobile communications equipment and devices. PTIC is leveraging its strong financial base and close connections with the Ministry of the Information Industry to shift the industry focus from pure manufacturing to technology and product development [Liu S. 1998b].

As a result, several foreign-Chinese joint ventures in the telecommunications industry have been established.

Datang Telecom Tech Group, a public company linked to the China Academy of Telecommunications Technologies, established a strategic alliance with Motorola to transfer mobile switching systems and GSM wireless phone system technologies. Zhongxing Telecom Ltd. (Shenzhen) established a joint DSP-technology laboratory with Texas Instruments to design wireless phones. Zhongxing is also collaborating with CommQuest Technology Ltd., an IBM subsidiary in China, to develop a GSM-based dual-band cellular phone. In addition, Huawei Technology Ltd. (Shenzhen) formed a three-thousand-person research team to develop telecommunications technologies such as ATM and SDH [Liu S. 1998b].

Foreign companies have also had to face smuggling, insufficient protection of intellectual property rights, unfair competition, and bureaucratic difficulties in China. Beijing, continues to issue contradictory regulations on telecommunications imports and foreign joint ventures. Foreign telecommunications companies could be hard hit if China enacts new joint-venture restrictions which target foreign joint ventures with Chinese partners that in turn provide financing to other Chinese entities to get orders[3] [Robertson 1998c]. However, such restrictions on foreign venture financing will not support China's case for admission to the World Trade Organization.

7.5   DEVELOPMENT GOALS FOR CHINA'S TELECOMMUNICATIONS INDUSTRY

The Chinese government's long-term strategy for the telecommunications industry includes installation of a nationwide fiber-optic trunk transmission network, and a multimedia communications network that combines public voice and data networks with mobile communications.

7.5.1   Goals for China's Telecommunications Industry

General development targets for China's telecommunications industry in the year 2000 include [*China IT Market Report* 1997]:

- increasing the rate of national telephone coverage to 9 to 10 percent, with the urban rate up to reach 30 to 40 percent,
- building the total capacity of the telephone network to 170 million lines, and the total capacity of local switches to nineteen million lines,
- expanding long distance automatic switch capacity to reach six million ports, and long-end business lines to reach 2.3 million,

---

[3] The so-called FCC (Foreign-Chinese-Chinese) model has been used extensively to sell to China Unicom. This model allows a separate Sino-foreign joint venture (F-C) to team with China Unicom (the second "C") to establish an operational entity [Harwit 1998].

- creating a long distance trunk line network to adopt primarily SDH and DXC equipment, local switches to fully realize digital program control, and to establish ISDN;
- developing an intelligence network using fiber-optic cable, N-ISDN business, ATM technology, and B-ISDN;
- in respect to mobile communication, realizing national networked roaming with the coexistence of GSM and an analogue system;
- appropriately adjusting network structure: the long distance telephone network will merge grade 1 and grade 2 switch centers and grade 3 and grade 4 switch centers; local networks will be built into integrated large urban and rural local networks, with prefecture and city-level central cities as the core and its own three long distance area codes;
- establishing three supporting networks consisting of a No. 7 signalizing network, SDH network, and network management system;
- focusing on the key technologies of the research: fiber-optic communications systems; SDH, GSM, CDMA, PCS (personal communications system), and B-ISDN technologies;

7.5.2   Assessments

The future of China's telecommunications industry looks promising. The market has been growing rapidly since the economic reform and opening to the outside world in the early 1980s. The growth potential is huge, due to continuous strong domestic demand for a variety of communications services. The telecommunications equipment market has been opened up to foreign investment and technology transfer. The level of technological sophistication and communications capacity has been increasing constantly over the years. However, China's telecommunications industry still faces many challenges in the years to come. They include:

- the lack of a complete set of laws and regulations to protect fair competition in the telecommunications market;
- contradictory regulations on telecommunications imports and foreign joint ventures;
- slowness to embrace advanced techniques for managing and regulating communications systems;
- restrictions on Internet access and political controls on its use;
- delayed expansion of service to rural areas.

To modernize its telecommunications infrastructure and management, China most further liberalize and open up the domestic market to promote fair competition and foreign investment. Laws and regulations regarding foreign joint ventures and subsidiaries should be consistent. Current restrictions on access to and the content of the Internet need to be lessened to facilitate information flow and the exchange of ideas. Modern management techniques in managing and regulating communication systems and networks must be adopted broadly. The country must also speed up the process of transferring foreign technology and expand efforts to develop indigenous technologies and manufacturing capability to produce equipment for both export and domestic use. Finally, China should actively participate in international telecommunications cooperation and standard-setting.

REFERENCES

AA Telecom Services, Inc. (AATS). "China telecomm information 1997." 1997. [On-line], Available: http://www.aats.com/new/telecom.htm

Advanced Semiconductor Manufacturing Corporation of Shanghai (ASMC). Company Website. June 1988. [On-line], Availabe: http://www.asmcs.com

Asia System Media Corp. "China's reform of the taxation system." 1997. [On-line], Available: http://www.cbw.com/business/chinatax

Baker, Peter and Helen Dewar "Clinton renews China's trade status; Cites nuclear issue." *Washington Post*, June 4, 1998.

Beijing Review (compilers). 1997. *China: Facts & Figures 1997*. Beijing: New Star Publishers.

Belling Microelectronics Manufacturing Co., Ltd. Company Website. October1996. [On-line], Available: http://203.94.0.109/belling.htm

Boulton, William. "Hong Kong –South China Electronics Industry." In M. Kelly and W. Boulton (eds.) *Electronics Manufacturing in the Pacific Rim*. Baltimore, MD: WTEC/Loyola College, 1997.

Carroll, Mark. "IT industry in China catching up to Taiwan's." *EE Times*. Business Section. 12 January 1998a.

Carroll, Mark. "Local vendors drive growth in China's PC arena." *EE Times*. News Section.6 April 1998b.

Carroll, Mark. "Downsizing of ministries off to a slow start in China." *EE Times*. 23 April 1998c.

China Computer Trends. *China Research Corporation*. [On-line], Available: http://www.china-research.com

China Economic Information Network (CEInet). [On-line], Available: http://www.chinaeco.com

China Electronics Industry Yearbook 1997. October 1997, Electronics Industry Publisher, Beijing, China (in Chinese).

China Electronics Information Center of the Ministry of the Electronics Industry (CEIEC). Homepage. 1997. [On-line], Available: http://www.ceic.gov.cn

China Information Technology (IT) Market Report, 1997. [On-line], Available: http://www.chinaeco.com/ecit.htm

China National Computer Software and Technology Service Corporation (CS&S). Homepage. 1998. [On-line] Available: http://www.ceic.gov.cn/enterprise/csia/cs.html

China National Network Center. The National Network Center, 1998. [On-line], Available: http://www.net.edu.cn

China Telecommunication Weekly. China Research Corporation. [On-line], Available: http://www.china-research.com

China Vista, "Lucent to complete contract." March 5, 1998. [On-line], Available: http://www.chinavista.com/business/news/archiev/mar98/mar05-02.html

Crisanti, Lynn. "Untangling China's datacom network." *China Business Review*, Vol. 24, Issue 5, September-October 1997, pp. 38-41.

Dreyer, June. T. *China's Political System: Modernization and Tradition,* 2nd ed. Allyn & Bacon. 1996.

Dunn, Darrell. "PCB makers seek foothold on slippery slope." *Electronic Buyers' News*, November 16, 1998.

Economist. "Ready to face the world? A survey of China." March 8, 1997.

Forbes, Jim. "PC makers take a slow boat to China." *Windows Magazine.* September 18, 1997.

Fu, Mandy. "PC giant Legend to go regional." *Channels Asia*, May 1998.

Haber, Lynn. "IBM offers E-business solutions in China." *Integration Management*, May 4, 1998.

Harwit, Eric. "China's telecommunications industry: Development patterns and policies." *Pacific Affairs*, Vol. 71, Issue 2, Summer 1998, pp. 175-193.

Hatori, M. *The report on the Strategy for the Electronic Circuits Industry.* Institute for Interconnecting and Packaging Electronics Circuits (IPC). Japan: The Japan Printed Circuit Association, 1996.

Howell, Thomas R., Jeffery D. Nuechterlein, and Susan B. Hester. *Semiconductors in China: Defining American Interests,* Semiconductor Industry Association, Dewey Ballantine, 1995.

IBM China. *IBM Information in China.* 1998. [On-line] Available: http://www.ibm.com.cn/ibmchina

IEEE. "Chip Making in China: Snapshots of China." *IEEE Spectrum,* December 1995.

Information Office of the State Council (IOSC) of the PRC. *Key Open Zones.* China International Press, August, 1997a.

Information Office of the State Council (IOSC). "Development of science and technology in China." *China International Press,* August, 1997b.

Institute of Software, Chinese Academy of Science (ISCAS). Homepage. 1998. [On-line], Available: http://www.ios.ac.cn/int/introduction.html

Intel Corporation. "Intel Commitment to China." Press Releases, May 1988. [On-line], Available: http://www.intel.se/apac/eng/andygrove/pr0505.htm

Jiang Zemin. "Hold high the great banner of Deng Xiaoping theory for an all-round advancement of the cause of building socialism with Chinese characteristics into the 21st century." Report delivered at the 15th National Congress of the Communist Party of China on September 12, 1997.

Koo, George P. "China seeks world-class semiconductor industry." *Channel Magazine.* November 1997. [On-line], Available: http://www.semi.org/Channel/1998/nov/ features/china.html

Krochmal, Mo. "China sees huge demand for Net content." TechWeb Technology News, May 14, 1998a. [On-line], Available: http://www.techweb.com

Krochmal, Mo. "China seeking filtered content." TechWeb Technology News, May 28, 1998b. [On-line], Available: http://www.techweb.com

Lammers, David. "Japan gaining inside track to China's chip market." *EE Times – Headline News* (CMP Media, Inc.), 1997. [On-line], Available: http://pubsys.cmp.com/eet/news/97/956news/japan.html

Lardy, Nicholas R. *China and the WTO*. Testimony before the House Committee on Ways and Means on the accession of China and Taiwan to the World Trade Organization. September 19, 1996.

Legend Computer Systems Ltd. Company Website. April 1998. [On-line], Available: http://www.webleader.com.cn/legend-pc

Li, Ning. "Moving toward the age of information – The rapid development of China's PC sector." *Beijing Review*, November 11, 1996.

Li, Wenfang. "Notebook computers dominate market." China Economic News. February 11, 1998. [On-line], Available: http://www.chinaeo.com/frame.htm

Liang, Xiongjian, Xueyuan Zhang, and Xu Yang. "The Development of Telecommunications in China." *IEEE Communications Magazines*, November 1998, pp. 54-58.

Liu, Sunray. "A design industry begins to grow up in China." *EE Times*, October 19, 1998a.

Liu, Sunray. "Emphasis switches to adding value to the network, ensuring service quality – China's telecom overhaul enters a new phase." *EE Times*. International Section. November 16, 1998b.

Liu, Xiaowen. "Competition boosts mobile telecommunications industry." *Beijing Review*, Vol. 41, No. 46, November 16, 1998, pp. 17-19.

Ma, Jun. *China's Economic Reform in the 1990s*. January 1997a. [On-line], Available: http://members.aol.com/junmanew/cover.htm

Ma, Jun. *Intergovernmental Relations and Economic Management in China*. New York, NY: MacMillan Press Ltd., 1997b.

Ministry of Electronics Industry (MEI). "Brief introduction to the Ministry of Electronics Industry." October 1997. [On-line], Available: http://www.ceic.gov.cn/mei/mei1.html

National Research Center for Intelligent Computing System (NCIC), Institute of Computing Technology, Chinese Academy of Sciences. Homepage. 1998. [On-line],   Available: http://www.solar.rtd.utk.edu/~china/ins/NCIC/NCIC.html

NEC. "NEC signs contract for China's largest semiconductor project." News Release, May 1997. [On-line], Available: http://www.nec.co.jp/english/today/newsrel /9705/2801.html

Niccolai, James. "IBM, China announce E-commerce pact." *ComputerWorld (Hong Kong).* June 8, 1998.

PRC. State Statistical Bureau, People's Republic of China, China Statistical Yearbook, China Statistical Publishing House, 1995.

The Office for Specialized Economic Zones (OSEZ) of the State Council of the PRC. "Main features of the economic & technological development zones in China." 1997. [On-line], Available: http://www.sezo.gov.cn/kfqtde.html

Qin, Shi. *China 1997.* Beijing, China: New Star Publishers, 1997.

Rehak, Alexandra and John Wang. "On the fast track – Modest liberalization in China's telecommunications sector means more sales and new competition." *China Business Review*, Vol. 23, Issue 2, March-April 1996, pp. 8-13.

Roberts, Dexter. "In China, No Great Wall across the Net." *BusinessWeek*, International edition, August 26, 1996a.

Roberts, Dexter. "The race to become China's Microsoft." *Business Week.* November 18, 1996b, p. 62.

Roberts, Dexter, and B. Einhorn. "Going toe to toe with Big Blue and Compaq."*Business Week 14* April 1997, p58.

Roberts, John, and Keith Burbank. "Chinese computer market – distribution as a percent of installed base of desktops, notebooks, and servers." *Computer Reseller News* Issue 808, September 21, 1998.

Robertson, Jack. "Dell joins rush to China." *Electronics Buyers' News*, September 11, 1997.

Robertson, Jack. "China's telecom shake-up: Now what?" *Electronic Buyers' News.* Issue 1106, News Section. April 27, 1998a.

Robertson, Jack. "Motorola plans "superfab" site in China." *Semiconductor Business News*, May 5, 1998b.

Robertson, Jack. "CCF model under attack." *Electronic Buyers' News*. Issue 1130, News Section. October 12, 1998c.

Rohwer, Jim and Stu Rohwer. *Asia Rising*. New York, NY: Simon & Schuster, 1996.

Schell, Orville. "Deng's revolution." *Newsweek*. March 3, 1997, pp.20-27.

Schumann, Elizabeth. "Market focus: China. A feature of SEMI's equipment and materials market statistics program." *Channel Magazine*, February 1997. [On-line], Available: http://www.semi.org/Channel/1997/feb/market.html

Shirk, Susan L. *The Political Logic of Economic Reform in China*. University of California Press, 1993.

Simon, Denis Fred. "From cold to hot – China struggles to protect and develop a world-class electronics industry." *The China Business Review*, Vol. 23, No. 6, November-December 1996, pp.8-16.

Southern Software Industry Park (SSIP) homepage. 1997. [On-line], Available: http://www.ceic.gov.cn/enterprise/csia/ssi.html

SPSP Development Company. Company website, 1998. [On-line], Available: http://www.ceic.gov.cn/enterprise/csia/pd.html

The State Council of China. *Decision on Accelerating Scientific and Technological Development*. National Conference on Science and Technology Policy, May 1995. Beijing, China.

Tsuda, Kenji. "China pushes submicron chip fabs." *Nikkei Electronics Asia*, Vol. 6, No. 2. February 1997. [On-line], Available: http://www.nikkeibp.com/nea/feb97/febchina.html

Ure, John. "Telecommunications in China and the four dragons." In Jone Ure (ed.) *Telecommunications in Asia*. Hong Kong: Hong Kong University Press, 1995, pp.11-48.

U.S. Department of Commerce (DOC), Office of Technology Policy, China: Strategy for technology acquisition. 1998. [Online], Available: http://www.ta.doc.gov/AsiaPac/china

U.S. Department of State (DOS). *1997 Country Reports on Economic Policy and Trade Practices: People's Republic of China*. Report submitted to the Senate Committees on Foreign Relations and on Finance and to the House Committees on Foreign Affairs and on Ways and Means, January 1998.

Viewsletter IT Weekly. New Century Group, Hong Kong, Issue 8, 1998. [On-line], Available: http://www.china-research.com

Wallace, Richard. "PC revolution sweeps China." *EE Times*, April 26, 1998.

Wang, James C.F. *Contemporary Chinese Politics: An Introduction*. Upper Saddle River, NJ: Prentice-Hall, 1980.

Weng, Shuo Song "A survey of the Chinese IC industry." *Solid State Technology China*, November 1996, 14-17. (in Chinese).

Wessel, David. "Big discrepancy exists between data from U.S. and China on trade deficit." *Wall Street Journal*, January 22, 1998.

Williams, Martyn. "NEC establishes chip design joint venture in China." *Newsbytes Pacifica*, March 16, 1998. [On-line], Available:http://www.nb-pacifica.com/healdine/necestablisheschipde_1262.html

Wong, Christine P.W., Christopher Heady, and Wing T. Woo. *Fiscal Management and Economic Reform in the People's Republic of China*. Hong Kong: Oxford University Press, 1995.

Wong, Y.C. Richard, and M.L. Sonia Wong. "Removing regulatory barriers in China: Changing the foreign exchange regime." Paper prepared for the conference of China as a global economic power: Market reforms in the new millennium, Shanghai, June 15-18, 1997. [On-line], Available: http://www.cato.org/events/china/papers/wong.html.

World Bank, *China Foreign Trade Reform*. Washington, D.C.: World Bank, 1994.

Yang, Tianxing. "The status and prospect of the development of China's software industry." China Electronics Information Center of the Ministry of the Electronics Industry, 1997. [On-line], Available: http://www.ceic.gov.cn

Zhang, Jeff X. and Y. Wang. *The Emerging Market of China's Computer Industry*. Westport, CT:Greenwood Publishing Group, 1995.

Index

Milton Keynes UK
Ingram Content Group UK Ltd.
UKHW031150141024
449569UK00024B/917